what makes the
WORLD
go round?

A Henry Holt Reference Book
Henry Holt and Company, Inc.
Publishers since 1866
115 West 18th Street
New York, New York 10011

Library of Congress Cataloging-in-Publication Data
What makes the world go round?: a question-and-answer encyclopedia /
Jinny Johnson, general editor.—1st ed.
p.   cm.—(Henry Holt reference book)
Includes index.
1. Science—Encyclopedias.   2. Encyclopedias and dictionaries—
Miscellanea.   3. Science—Miscellanea.   I. Johnson, Jinny.
I. Series.
Q121.W48   1997   96–41111
503—dc20   CIP

ISBN 0-8050-5086-8

First American Edition—1997
This book was conceived, edited, and designed by Marshall Editions
170 Piccadilly, London W1V 9DD

Art Editor: Smiljka Surla
Design Manager: Ralph Pitchford
Managing Editor: Kate Phelps
Picture Editor: Elizabeth Loving
Copy Editor: Jolika Feszt
Art Director: Branka Surla
Editorial Director: Cynthia O'Brien
Production: Janice Storr, Selby Sinton

Printed and bound in Italy
Originated in Italy by Ad Ver Srl
All first editions are printed on acid-free paper.∞

10 9 8 7 6 5 4 3 2 1

The publishers would like to thank the Natural History Museum, London,
and Joyce Pope for the use of text from *Do Animals Dream?* on pp. 38–39,
40–41, 42, 45, 50–51, 52–53, 54–55, and 60–61 of this book.

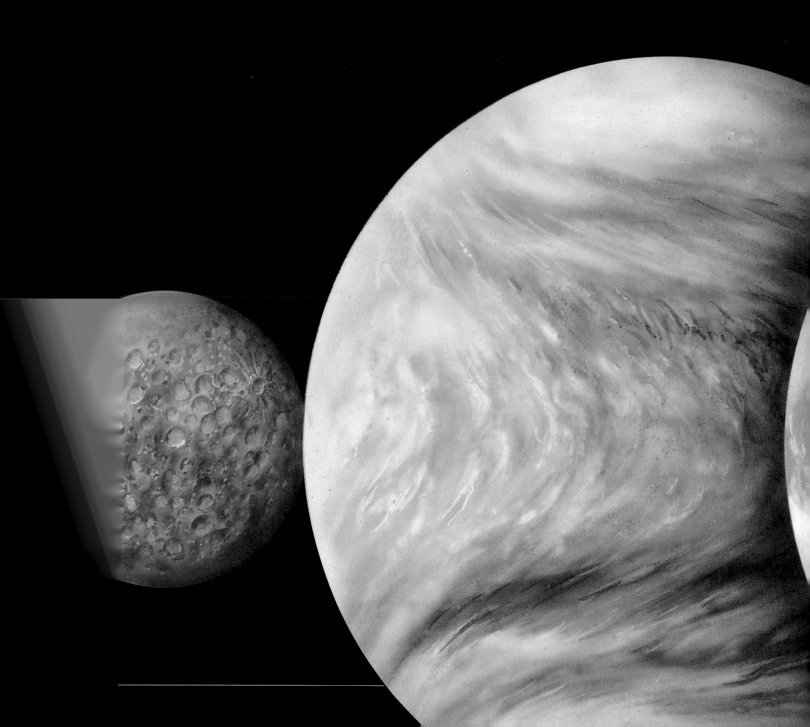

what makes the
WORLD
go round?

A Question-and-Answer Encyclopedia

General Editor
Jinny Johnson

Consultant Editor
Dr. Philip Whitfield

A Henry Holt
Reference Book

Henry Holt and Company
New York

Contents

8 The World of the Past

How did life begin?
What are fossils?
What were the first forests like?
Were all dinosaurs big and fierce?
What were pterodactyls?
Why did the dinosaurs die out?
Did the first people look like us?
Where did the first humans live?
What were the first farmed animals?

And more than 25 other questions and answers
on early life, dinosaurs, and the first humans.

32 The Living World

Do all birds sing?
How do whales breathe?
Why do tigers have stripes?
Which are the most dangerous animals?
Why do plants have flowers?
Where are the world's deserts?
What animals live in a rain forest?
What is tundra?
Why do creatures die out?

And more than 70 other
questions and answers on animals,
plants, and natural habitats.

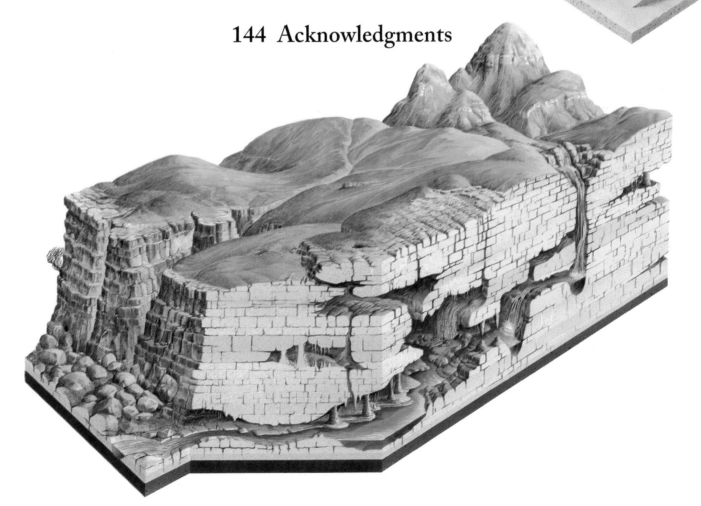

Foreword

Have you ever wondered how life began, why birds migrate, what coral reefs are made of, why plants have flowers, why the sea has tides? The world around us is a fascinating and incredible place, filled with such mysteries, many of which this book helps to explain.

The first chapter looks at questions about the past—early life, dinosaurs, the first plants, and the first people. In these pages we travel back in time to look at some of the earliest history of our planet.

The second chapter concentrates on the miracles of life on our planet today—hovering hummingbirds, insects that look like leaves, the biggest flowers. It also examines some of the world's main habitats, such as rain forests, deserts, and tundra, and the creatures and plants that live in them.

The third chapter explains the workings of the Earth itself—questions such as where rivers begin, how a glacier moves, what causes a tornado, what is under the ground?

In tackling a wide variety of questions about the origins, life, and workings of our planet, this book provides a wealth of stimulating and entertaining information.

The world of the past

The story of the past is much more than a series of dates. It is a fascinating tale of exciting developments that created the world of today and will continue to shape the future.

Once life had begun on Earth, the planet changed forever, as living things began their amazing, never-ending game of variations called evolution. Scientists know about some of these early creatures from their fossilized remains, discovered in rock. Most remarkable of all the animals that roamed the ancient Earth were the dinosaurs, which ruled the planet for more than 100 million years before they disappeared in one of the great mysteries of the past. They included the largest land animals known to have lived in this world.

It was not until much later, about three million years ago, that our apelike ancestors first walked the Earth and began the long process of evolving into humans. Gradually, they populated the world and started to make tools, grow food, and keep animals. A journey back in time poses intriguing questions about the past. In the following pages, the answers to many of these bring these ancient worlds vividly back to life.

A school of fossilized fish (left), which lived in the Eocene period, 54 to 38 million years ago. Unusually well-preserved fossils, they show even the scales on the bodies of the fish.

How did life begin?

This is a question that no scientist can answer completely. Most scientists, though, have some ideas about when life first started on Earth and how it was organized. No life at all was possible on Earth before there were seas. However, less than a billion years after the solid crust first formed, there were seas containing minute forms of life. Microscopic fossils are preserved in seabed rocks that are more than 3.5 billion years old. These fossils seem to have been sea-dwelling creatures, each made up of a living unit—a single cell. They were like the microscopic organisms called bacteria that we know today. Since there was no oxygen in the air or seas at that time, these early living things, unlike most living things today, must have been able to live without oxygen.

So life must have had its beginnings at some point between the formation of the seas and the first appearance of simple cells. It is thought that life developed as clusters of molecules in the sea, all containing the element carbon. Gradually, these molecules became more complex, until they were able to make copies of themselves. Once they could do that they were really alive, and the evolution of life could begin.

By 3.2 billion years ago, the first plantlike bacteria, called blue-green algae, had evolved. These bacteria could trap sunlight and carbon dioxide from the air and use them to make new living material in the process called photosynthesis. A waste product of photosynthesis is oxygen, and the levels of oxygen in the Earth's air began to increase. This change probably triggered the evolution of a variety of new, more complex living things, including plants and animals. These new animals needed oxygen to live and breathe, and they gave off carbon dioxide as waste. These animals could only have appeared on Earth after the plants had released enough oxygen into the atmosphere.

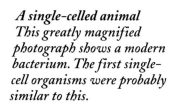

*A single-celled animal
This greatly magnified photograph shows a modern bacterium. The first single-cell organisms were probably similar to this.*

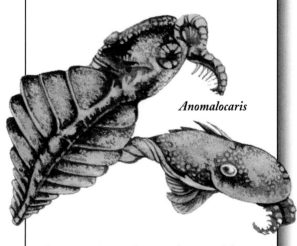

Armored lobopod

The armored lobopod was one of the strangest animals of 500 million years ago. This wormlike creature had many pairs of legs tipped with claws. On the top of the body were hard plates and spines which protected it from enemies.

Habelia

Habelia was an early arthropod from the Burgess Shale. Its head bore short antennae. Behind that were six pairs of branched legs with which it was able to walk in the mud.

Anomalocaris

Anomalocaris was the great hunter of the Cambrian seas and was much larger than most Burgess Shale creatures. This 2-foot-long arthropod navigated with its huge eyes and grasped its prey with two large arms at the front of its body.

What were the earliest animals like?

The many creatures that lived nearly a billion years ago in the Precambrian time included true animals. They moved about, ate, and had young. But their bodies were so soft that they quickly rotted after death. Fossils are usually formed from the hard parts of an animal, so no one really knows what these creatures looked like.

The earliest animals that can be described in detail come from a time called the Cambrian period, around 500 million years ago, when the sea was bursting with new animal life. Many of these animals grew shells, scales, or other types of protective body armor. These hard parts, because they could be preserved in the rock, produced the first fossils to show any detail of the creatures' bodies.

Cambrian life included examples of almost all the main types of sea animals without backbones that live today. There were sponges, corals, worms, and many types of arthropods—animals with a skeleton on the outside, jointed limbs, and a body made up of segments. Today's arthropods include creatures such as insects, spiders, crabs, and shrimp.

Where can their remains be seen?

The remains of these creatures can be found in many places on Earth. Recent finds have been made in Greenland and China. Some of the best-known such fossils are those found in a quarry known as the Burgess Shale in the Canadian Rockies.

These fossils were first discovered in 1909 by Charles Dolittle Walcott, then America's most famous fossil-hunter. He collected thousands of specimens and began describing them. But he assumed that these animals belonged to the same main groups that exist today. He missed the fact that the Burgess Shale animals included not only most modern groups without backbones, but also creatures which have no living lookalikes.

What are fossils?

Fossils are the remains or traces of dead animals that have been preserved in sedimentary rocks. (Sedimentary rocks are formed by slowly accumulating sand and mud.) From a fossil, scientists can get clues about the climate and geography of the time as well as information about how, when, and where that particular animal lived.

Fossils are usually made from some hard part of the living thing which did not rot away, such as a shell, bones, teeth, or scales. While the fleshy parts of the organism gradually decayed, the hard parts were preserved in sand, mud, or clay. Over the centuries, these harder materials slowly turned into some type of rock and became entombed.

Trace fossils such as dinosaur footprints in rock or burrows made by soft-bodied creatures also give clues about the past. Dinosaurs' footprints, for example, were originally made in soft mud. The mud in which the footprint was made dried and hardened before new layers of sediment covered up the track, preserving it forever.

Can plants become fossils?

Stromatolites at Shark Bay, Australia

Yes. All types of plants, even microscopic ones, such as single-celled algae, can form fossils if conditions are right at the time of their death.

Some of the oldest fossils in existence were formed by primitive plants. These blue-green algae lived in shallow seas and were among the first life forms that could make energy from sunlight by the process of photosynthesis. A mass of tiny filaments, they trapped particles of limy minerals, eventually forming large mounds. These mounds are preserved as fossils called stromatolites.

Land plants with fronds, leaves, seeds, fruits, and wood have also left many fossils. Where those remains fell into mud without oxygen, the plant remnants did not rot easily, and perfect fossils resulted.

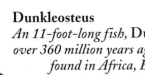

Dunkleosteus
An 11-foot-long fish, Dunkleosteus *lived over 360 million years ago. Fossils have been found in Africa, Europe, and North America. A large and powerful hunter, it chopped at its fish prey with sharp dental plates—not separate teeth.*

Bones or scales may be turned into rock, forming fossils. An animal or plant that is not preserved as a fossil may still leave an impression or a trace either as a cast (top left) or a mold (bottom left). These are called trace fossils. Here, the shell of an ammonite has been fossilized, leaving impressions in the rock lying above and below.

How old are fossils?

The oldest fossils so far discovered are more than 3 billion years old. It is rare to find fossils this old for a number of reasons. First, the life forms of that time consisted of microscopic organisms similar to today's bacteria, and to observe life this small, a very powerful microscope is needed. Second, fossils form in sedimentary rocks, and the older these are, the greater the chance of fossil remains being destroyed. They can be worn away by the forces of the sea or the weather, or changed by heat and pressure deep underground.

The oldest known sedimentary rocks are found in southern Africa and are more than 3 billion years old. When these rocks are cut into slices thin enough for light to pass through and examined under a microscope, the fossilized remains of minute creatures can be clearly seen.

Fossils exist in most types of sedimentary rocks. Because scientists can find out how old these rocks are, the fossils entombed in them have helped them to form a picture of the early plant and animal life on Earth.

In the rocks of the Cambrian period, formed between 500 and 600 million years ago, other small fossils may be found, including fossil trilobites (creatures similar in appearance to modern woodlice) and brachiopods (lampshells). In the succeeding Ordovician period, creatures such as sea urchins and starfish appeared. Early fossil fish are found in rocks about 400 million years old from the Devonian period, reptile fossils have been discovered in rocks dating from about 300 million years ago, and the first true mammals have been found in Jurassic rocks about 200 million years old.

Ancient mollusks
Ammonites were fossil mollusks that are now extinct. They date from the Jurassic period, about 200 million years ago.

EARLY PLANTS

During the 150 million years following the Carboniferous—the Permian, Triassic, and Jurassic periods—plants such as seed ferns, cycads, horsetails, and club moss still dominated the Earth. Palmlike cycads were at their most varied and abundant in the Jurassic. Flowering plants did not really begin to thrive until the Cretaceous period, which began about 135 million years ago.

Horsetail

Cycad

Club moss

Seed fern

What were the first forests like?

The first forests were dominated by tree-sized fernlike plants that grew to about 100 feet high. The first plants to grow on land at all, however, were small and simple. They formed meadows, rather than forests. From them, more complex and much larger plants began to develop. During the Devonian period, about 400 to 360 million years ago, small ferns and then the more massive fernlike trees appeared.

More mixed forests, with plants of many different heights, evolved in the Carboniferous period, between 350 and 270 million years ago, when the world's weather was warm and damp. The ancient club moss *Lepidodendron* and the horsetail *Calamites* reached up to 90 feet into the sky. Today, plants of these types grow to 3 feet at the most. Other trees included the ancestors of today's gingko tree, monkey puzzles, and cycads.

What animals lived in the forests?

The moist forests of the Carboniferous period were inhabited by a wide range of animals. These included fish, amphibians, and even reptiles, as well as many different insects, spiders, and scorpions. The forests covered much of what is now North

Ancient forests
In the dense, swampy forests of the Carboniferous period, giant dragonflies and other insects flew among the towering tree ferns, club mosses, and horsetails.

America and Europe, which at that time were joined together. The fossils found there tell us that springtail insects jumped among the rotting vegetation, millipedes munched their way through it, and centipedes hunted for worms and other invertebrates, which they paralyzed with their poisonous fangs.

Among the many early insects which flew about between the trees, the most massive were dragonflies, with a wingspan of up to 27 inches—the giants of the Carboniferous sky. Amphibians varied in size from those the size and shape of today's salamanders to enormous, crocodile-sized predators.

What happened to this forest life when it died?

Many of the plants and animals that died in the Carboniferous forests were eaten by other animals or broken down by the tiny microbes which cause decay. Many others were turned into fossils which have given us a good idea of what the forests looked like. A lot of the fossilized, often partly rotted, plant material built up as peat. This eventually turned into coal. The name of this period—the Carboniferous—means "coal bearing."

Dead plants that fell into the swamp's mud were often buried before they could decay very much. Some of this mud had so little oxygen in it that microbes were virtually unable to decompose plants and tree trunks.

Over millions of years, thick layers of this plant material built up as peat, which was later covered by layers of mud. When this mud was pressed down by the weight of the layers above it and slowly turned into rock, the plant material itself slowly altered and, in the end, turned into the black substance we call coal.

A fossilized fern frond

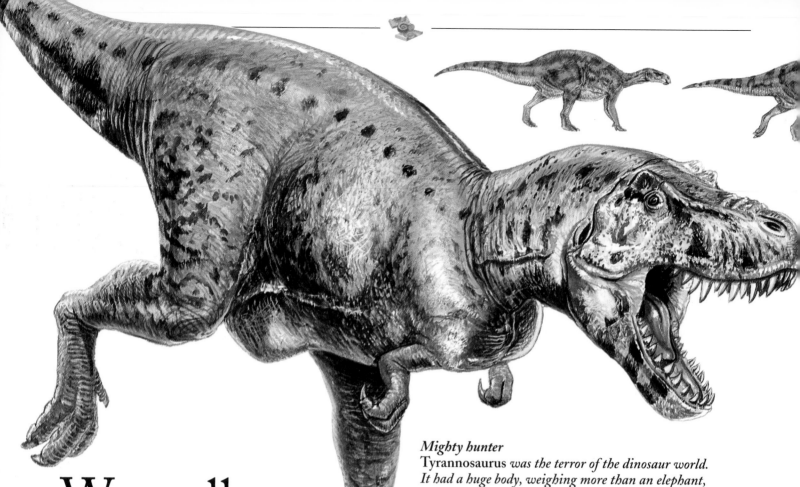

Were all dinosaurs big and fierce?

Mighty hunter
Tyrannosaurus *was the terror of the dinosaur world. It had a huge body, weighing more than an elephant, and its massive head was filled with big, razor-sharp teeth shaped like steak knives, with serrations down their sides. Its powerful jaws would have been big enough to swallow a person whole.*

The name dinosaur means "terrible lizard," but by no means all these creatures were big and fierce. While some dinosaurs were as tall as a three-story building or as heavy as a 30-ton whale, others were only the size of a chicken. And although certain dinosaurs had sharp claws and vicious teeth for tearing apart the flesh of their prey, others, such as *Iguanodon*, were peaceful plant-eaters and grazed quietly in herds or browsed among the treetops. These plant-eaters were preyed on by the meat-eating dinosaurs.

Tyrannosaurus, one of the best known of all the dinosaurs, was certainly big and fierce, however. This giant predator, whose name means "tyrant lizard," was the biggest hunting animal ever known on Earth. It lived about 70 million years ago in the Late Cretaceous,

toward the end of the dinosaur era.
About 12 feet tall, it measured nearly 40 feet from head to tail. Its mouth was lined with 60 daggerlike teeth, some up to 6 inches in length, and its hind legs had sharp claws on the toes. With these monstrous jaws and teeth, this dinosaur could kill and rip prey apart in a matter of seconds.

Tyrannosaurus is the most fearsome example of one type of meat-eating dinosaur, the carnosaurs, which were all big and powerful animals. But other hunters, such as *Gallimimus* and *Ornitholestes*, were lightly built, fast-running dinosaurs, which fed on small prey, such as mammals, lizards, and insects. Speed was their best defense against danger. They did not have massive teeth and claws, but few other dinosaurs could catch them when they ran at full speed.

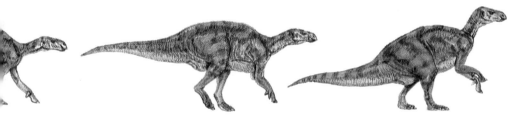

What color were dinosaurs?

No one knows. The color of an animal's skin is one of the first things to be lost after it dies, even if, in time, it turns into a fossil. The chemicals that make up some of the colors fade and then rot away soon after the animal has stopped living.

Even when the tough skin of a dinosaur has been preserved as a fossil, it does not show any of the colors or markings that it would have had when alive. Its color will be that of the mineral in the fossil rock.

When scientists and artists create illustrations of dinosaurs, they have to guess what colors and patterns to choose. These guesses are based partly on the variety of colors and patterns seen in today's reptiles and partly on what is known about the way of life and habitats of the different dinosaurs.

Hunting in packs
Groups of **Gallimimus** *sped across plains, searching for small prey such as lizards. They also pulled down the branches of trees with their long, clawed fingers, and fed on leaves, buds, and fruits.*

Were dinosaurs warm-blooded?

For many years biologists who studied the fossilized remains of dinosaurs thought that these animals must all have been cold-blooded, like today's reptiles and amphibians. They believed that they would have been unable to control their temperature and keep it high like warm-blooded mammals do today.

Some fossil experts now think that certain of the larger dinosaurs might have been warm-blooded. But keeping a constant body temperature uses up a lot of food. It is difficult to see how some of the huge plant-eating dinosaurs could have eaten enough to provide the energy to keep their vast bodies at a high temperature, especially since they lacked good insulation such as hair or feathers.

Some large dinosaurs may have kept a high body temperature partly by making energy from their food and partly by their lifestyle, for instance moving in and out of the sun to keep them warm. Another idea is that the large plates on the backs of dinosaurs such as *Stegosaurus* could have been thermal panels, picking up solar heat when facing toward the sun. We will probably never know the truth.

Fast mover
Ornitholestes *was a swift-moving hunter. Its name means "bird robber," and it may have preyed on early birds or flying insects as well as small reptiles.*

Which was the biggest dinosaur?

The biggest dinosaurs were the giant, long-necked plant-eaters known as sauropods, some of which may have been more than 100 feet long. These sauropods, which included animals such as *Brachiosaurus* and *Diplodocus*, were the largest land animals ever to have lived on Earth.

Diplodocus may have reached a length of 100 feet, but it weighed only about 11 tons. The special structure of its vertebrae (the bones that form the backbone) made *Diplodocus* light for its size. The vertebrae were partly hollow, so they weighed less than if they were made of solid bone. *Brachiosaurus* was only about 75 feet long, but it had a much heavier body. It may have weighed up to 89 tons—as much as 12 African elephants today.

A few bones which may belong to even larger sauropods have been unearthed in recent years. One creature, at present called *Seismosaurus*, may have been more than 130 feet long.

How many sorts of dinosaurs were there?

From the fossils found to date, there are thought to have been more than 400, from gigantic animals plodding along on all fours down to agile, fast-running birdlike dinosaurs. There were meat-eating carnivores, plant-eating herbivores, and omnivores, not fussy in their feeding habits.

Scientists divide the dinosaurs into two main groups—the lizard-hipped and the bird-hipped dinosaurs—according to the design of their skeletons. As the name suggests, the most important differences have to do with the design of the hips, and the way the hip bones have been changed for walking on the hind legs only.

The lizard-hipped types of dinosaurs did not in fact all walk on two legs. In general, the meat-eaters did, while the plant-eaters supported their heavy bodies on four stout limbs. *Tyrannosaurus* was a two-legged lizard-hipped carnivore, while the even more massive *Brachiosaurus* and *Diplodocus* were typical lizard-hipped herbivores. All the bird-hipped dinosaurs were herbivores, with jaws perfectly adapted for eating plants.

The greatest number of dinosaur species have been found in rock from the Cretaceous period, when flowering plants first evolved. This new food source may well have led to the evolution of many new types of plant-eating dinosaur.

Has any human being ever seen a dinosaur?

The answer to this question is no. By the time that people first walked on the face of the Earth, the last dinosaur had been extinct for a very long time—more than 60 million years. So neither our own species nor our close ape ancestors has ever seen or come into contact with a living dinosaur.

The pictures of dinosaurs and the full-size models seen in some museums are all based on the fossilized remains found in rocks around the world. Using this information scientists have worked out what these amazing animals may have looked like.

In the novel and film *Jurassic Park,* the paleontologists' dream came true, and living dinosaurs were created from fossilized genetic material. Unfortunately, this exciting idea is not really possible. The only moving dinosaurs we will ever see are robotic models.

Where are dinosaurs found?

Fossil-hunters are still finding new types of dinosaurs in rocks all over the world formed in Triassic, Jurassic, and Cretaceous times. The first fossilized dinosaur bones and teeth were collected in the early 19th century. They came from the rocks of southern England and were given their first scientific names in 1824 and 1825. It was not until 1841 that Sir Richard Owen, who later became director of the Natural History Museum in London, realized that these strange and gigantic animals belonged to a completely separate and so far undiscovered group of reptiles. He gave them the name dinosaur, meaning "terrible lizard."

When the dinosaurs first came into the world at the end of the Triassic period, some 200 million years ago, all the landmasses of the planet were connected together in a super-continent called Pangaea. This meant that land animals, including the early dinosaurs, could roam to all regions of the world without having to cross seas.

Throughout the Jurassic period and until the later part of the Cretaceous period, all types of dinosaurs lived all over the world. Once Pangaea began to break up, and with the opening of the Atlantic Ocean to the west of Africa, fossils show that different types of dinosaurs developed in particular parts of the world. The greatest number of different dinosaur types has been found in North America, Europe, and Asia.

Are new dinosaurs still found today?

Yes, completely new groups of dinosaurs are still being discovered. One of the most recent finds is a new family of meat-eating dinosaurs from the early part of the Cretaceous period. The only member of this family found so far was discovered by an amateur fossil-hunter in a brick-clay pit in southern England in 1983. The dinosaur has been scientifically named *Baryonyx walkeri,* for Mr. Walker who discovered it, but it has since been given the more catchy nickname of "Claws."

Claws was a carnivore, 26–32 feet long with a crocodile-shaped head filled with many sharp, pointed teeth. On each of its forefeet was a huge curved claw about 12 inches long, which it used to attack its prey.

In the region where the creature's intestines would have been, the *Baryonyx* fossil contained scales and teeth from a large, lagoon-living fish. Scientists think that this dinosaur lived close to the water's edge and got its meals like grizzly bears do today, by hooking fish out of the water with its claws.

Did dinosaurs lay eggs?

Yes. Like all reptiles, dinosaurs did lay eggs. The reptile egg was one of the secrets of the dinosaurs' success. Amphibian eggs have only a coating of water and have to be laid in water so that they do not dry out. A reptile egg has a hard or leathery shell. This keeps the developing embryo from drying out and protects it from predators. Safe inside the egg, the reptile can grow until it is big enough to survive on land.

What were pterodactyls?

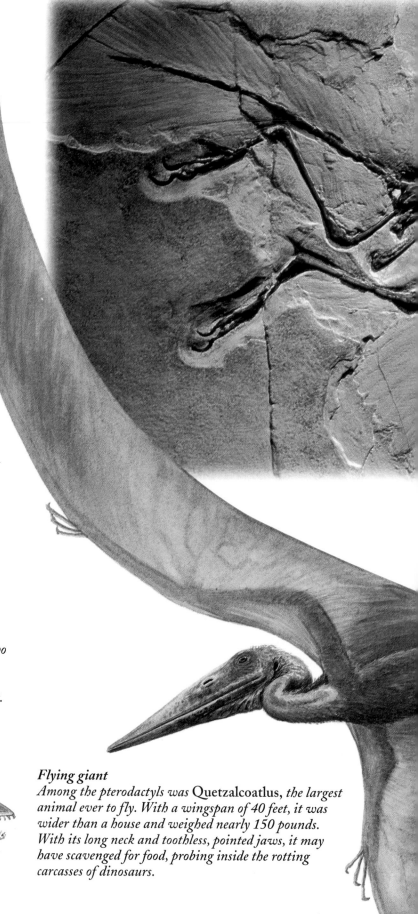

Pterodactyls were prehistoric flying and gliding reptiles related to dinosaurs. They belonged to the group of winged reptiles called pterosaurs, or "winged lizards," which first lived more than 200 million years ago—50 million years before the first known bird.

As far as is known, pterosaurs were the first vertebrates—animals with backbones—to fly. They had wings made of skin, which stretched between one enormously long "finger" on each hand and the back legs, and were attached to the sides of the body. The remaining "fingers" of each hand were short claws at the front edge of the wing.

There were two kinds of pterosaurs. The early forms, called rhamphorhynchs, had short legs and long, bony tails. Later came the pterodactyls, which had short tails and longer necks and legs.

Walking pterosaur
Experts disagree about how pterosaurs such as Anhanguera moved on the ground. Some believe that they walked upright on the two back feet like a bird. Others think that they crawled along using the claws on their front wings as well as their feet.

Flying giant
Among the pterodactyls was Quetzalcoatlus, *the largest animal ever to fly. With a wingspan of 40 feet, it was wider than a house and weighed nearly 150 pounds. With its long neck and toothless, pointed jaws, it may have scavenged for food, probing inside the rotting carcasses of dinosaurs.*

Fossil discovery
The fossilized remains of Archaeopteryx *(left), the earliest
known bird, were first discovered in southern Germany in
1861. Because they were preserved in fine-grained
sediments, plenty of detail can be seen, including
the feathers of the wings and tail.*

Were there birds in the sky when the dinosaurs lived?

Yes, there were. The earliest true
birds were warm-blooded,
backboned animals. They were quite
different from the winged pterosaurs
because they had feathers. Birds evolved
from small, meat-eating dinosaurs and
shared their world, flying in the same
skies as the pterosaurs. But when the
last of the flying reptiles became
extinct at the end of the Cretaceous
period, the feathered birds carried on. There
are about 9,000 kinds in the world today.

Nearly everything we know about the
earliest birds comes from some detailed
well-preserved fossils dating from about
150 million years ago. They are of the earliest
known bird, called *Archaeopteryx*, which was
15 inches long, about the size of a large
pigeon. Although it was like a reptile in many
ways—it had sharp, pointed teeth, claws on
its wings, and a long, bony tail—it was
undoubtedly a bird that flew, because the
fossils show both wings and feathers.

Although *Archaeopteryx* had a wishbone
like modern birds, it did not have their huge
breastbone to which the flight muscles are
attached. This suggests that *Archaeopteryx* was
not a strong flier and probably had to launch
itself into the air from a tree.

The first bird
A reconstruction of **Archaeopteryx** *shows its large wings,
long legs, and tail, which had bones down its center.
Scientists think it may have used its clawed wings and feet
to climb up the trunks of a tree and then flapped or glided
from one tree to another.*

Why did the dinosaurs die out?

Scientists are not really sure what made the last of the species of dinosaurs die out 65 million years ago at the end of the Cretaceous period. But they have several interesting theories about what might have happened.

Probably the most generally accepted idea is that the Earth at that time suffered a collision with a huge asteroid or meteorite. A big enough object falling to Earth from space would have produced an explosive force and damage similar to that of an all-out nuclear war. Vast amounts of dust, steam, and smoke pushed into the atmosphere would have blocked out the Sun's heat and light for a very long time. This lack of light could have killed off many land plants, and

Carnivorous dinosaur
Dromaeosaurus *lived about 70 million years ago during the late Cretaceous. These fierce predators hunted in packs.*

so caused the deaths of both the plant-eating dinosaurs and the carnivorous dinosaurs that fed on them. If the climate cooled down dramatically because of the dust in the atmosphere, this could also have contributed to the extinction of dinosaurs adapted for warm conditions.

There seems to be good evidence for this "space invader" explanation of the mass extinctions. In the rocks formed 65 million years ago, there is a thin band containing metals such as osmium and iridium that are rare in ordinary Earth rocks but more common in some types of meteorite. Perhaps the metals arrived from space. The band also contains tiny shattered particles of glass, made from fused sand grains, like those formed in intense explosions, and fossilized soot particles which may have come from forest fires caused by the fireball of a large meteorite impact. Studies have shown that the likely site of such an impact is a huge crater discovered off the Yucatán coast on the south side of the Gulf of Mexico.

Plant-eater
Lambeosaurus *was a plant-eating dinosaur. It lived in North America about 76 million years ago, toward the end of the dinosaur era.*

Did other creatures die out at the same time?

Yes, they did. When talking of the extinction of the dinosaurs, it is easy to forget that on the land and in the sea, many other large groups of creatures were dying out at the same time. The fact that both marine and land animals suffered together suggests that the cause must have been large enough to affect both the oceans and dry land equally. On land, all of the flying and gliding reptiles called pterosaurs (see page 18) disappeared at almost exactly the same time as their dinosaur cousins. In the seas and oceans, the swimming reptiles such as plesiosaurs and ichthyosaurs became extinct. So, too, did the belemnites and ammonites—two ancient groups of relatives of octopus and squid. Even many species of the tiny creatures in the sea called plankton were lost.

This disappearance of such a wide range of types of animals at one time is called a "mass extinction." The mass extinction at the end of the Cretaceous period set the scene for the development and evolution of a modern mixture of animal types. The major groups of animals that survived the mass extinction 65 million years ago were the direct ancestors of today's familiar animals—mammals, birds, lizards, crocodiles, turtles, fish, snails, and shellfish.

Their ancestors drew the lucky numbers in the lottery 65 million years ago and were able to take advantage of the vacant niches left by the disappearance of the dinosaurs and other creatures. Early mammals, for example, may have been able to control their body temperature more efficiently than dinosaurs could and to make use of underground habitats. This would have helped them survive a worldwide catastrophe.

SEA REPTILES

Ichthyosaurs lived much like the dolphins of today. Fast and agile, these marine reptiles could speed through the sea, hunting prey such as fish and squid. The biggest ichthyosaur ever found was nearly 50 feet long, but most measured only 6–13 feet. Ichthyosaurs became extinct at the end of the Cretaceous period.

Plesiosaur
Long-necked marine plesiosaurs, such as Elasmosaurus, *also died out with the dinosaurs.*

Pterosaur
Flying reptiles, such as Pteranodon, *disappeared at the end of the Cretaceous period, at the same time as the dinosaurs.*

Did early mammals lay eggs?

At least one group of primitive mammals almost certainly did lay eggs, like their reptile ancestors. These egg-laying creatures were ancient relatives of the modern animals called monotremes. The monotremes include echidnas, or spiny anteaters, and the duckbilled platypus, all found in Australia.

These mammals have fur and milk-making mammary glands, but unlike other modern mammals, they also lay eggs. The ancestors of today's monotremes were small animals which ranged in size from that of a shrew to a squirrel. They lived in a world ruled by the dinosaurs.

Scientists cannot be sure, however, that these creatures did lay eggs, because there are no fossils. Any such eggs would have been soft-shelled and rarely fossilized. Also, it is not certain whether all groups of modern mammals evolved from the same ancient creatures as the monotremes. The mammals of today that do not lay eggs could have come from a quite different set of prehistoric mammal ancestors which had already stopped laying eggs.

Did mammals evolve from dinosaurs?

The ancestors of the mammals were ancient reptiles, but not dinosaurs. The great difference between mammals and reptiles is that mammals are warm-blooded—they can keep their bodies warm all the time. Instead of depending on the sun to warm them up, mammals eat large amounts of food and burn it up quickly to turn it into heat.

This new method of controlling body temperature by eating more food required better jaws and teeth, and a more efficient way of walking. The mammals also needed to keep the heat in their bodies—the main way in which mammals conserve heat is by their covering of hair. Another distinguishing mark of a mammal is the one which gives the group its name—female mammals have breasts (mammary glands), which produce milk to feed the young until they can feed themselves.

The earliest mammal ancestors, a group of "mammal-like reptiles" called pelycosaurs which roamed the Earth 300 million years ago, are known only from fossils. Some pelycosaurs could control their body temperature by means of heat-regulating "sails" on their backs. The sails allowed the creature to absorb heat from the Sun, or to cool down, as needed.

Therapsids, the advanced mammal-like reptiles which appeared in the Triassic period, about 250 million years ago, were the direct ancestors of mammals. Only one group of these survived into the Jurassic period, 195 to 135 million years ago.

Did humans really evolve from apes?

Yes, they did. Just as mammals and birds arose from reptiles that existed before them, and amphibians evolved from fish, we too arose from creatures that lived on Earth before us. If we look around us in the animal kingdom for clues about who our ancestors might have been, the answer is obvious. There are very few other mammals that can walk on two legs, have virtually no tail, faces without a long snout, eyes on the front of the head that look straight ahead, and big brains. Apart from ourselves, only the apes have all of these features.

Scientists can now look at the genes of different animals—the inherited material that makes each animal what it is—and compare them. When the genes of people and chimpanzees are compared, it turns out that almost all our genes are exactly the same. This proves beyond any doubt that we are closely related to apes. Our closest living relatives among other animals are gorillas, chimpanzees, and orangutans. We and these living apes must have evolved from earlier forms of ape.

Are there fossils of early humans?

There are fossils of early humans—fossils of early members of our species and of the more apelike humans with whom we share distant ancestors. Unfortunately, there are very few complete ones.

One of the most famous human fossils found is from the "apewoman" nicknamed Lucy. She lived three million years ago in what is now Ethiopia. Lucy was not a member of the modern human species, known as *Homo sapiens*. She belonged to a now extinct species called *Australopithecus afarensis*, which was an early branch of our complex family tree. The fossils were found in 1974 by a team of American scientists.

What is an ice age?

An ice age is a period of time when the Earth gets colder all over. At times during an ice age, the temperature drops low enough to make the icy regions at the North and South poles get larger.

During the last 600 million years, there have been three major ice ages, each lasting millions of years. Within each ice age, there are regular changes between colder and warmer periods. In the colder times, much more of the world is covered with snow and ice than it is at present. The ice caps at the North and South poles get larger. Sea levels drop, and the weather is drier because so much water is locked up as ice.

In warmer periods, known as interglacials, the glaciers shrink, sea levels rise, and the weather gets wetter. The current Ice Age began about two million years ago and has since swung regularly between warm and cold periods. We are now in a warm, wet spell of the current Ice Age.

How do animals survive an ice age?

Animals survive the cold of an ice age by moving out of reach of the ice caps or by adapting to the colder conditions. The coldest periods of ice ages do not happen overnight.

During the lifetime of any one creature, even a long-lived animal such as an elephant, changes in climate are far too small to be noticed. Plants and animals have plenty of time to adapt to the different conditions. For example, when northern Europe was last covered with ice sheets 20,000 years ago, insect-eating birds such as warblers moved farther south to find warmer homes and plentiful supplies of their insect prey in the Mediterranean area.

Other animals gradually adapted to the cold. Elephants and rhinoceros, which today live in hot countries, managed surprisingly well. Special kinds evolved with long hairy coats to keep them warm. These wooly mammoths and rhinoceros flourished during the coldest times of the ice age.

Did the first people look like us?

Although they still had many apelike features, early people also had some **human characteristics.** Fossil experts are skilled at working out what an animal looks like from fossil bones. Their reconstructions have been used to produce "identikit" pictures of what we think the early man-apes and humans looked like.

Some kinds of man-ape were much shorter than a modern person, being only about 3–4 feet tall and weighing about 65 pounds. In many ways, they looked like upright chimpanzees. They were probably hairy, and their faces resembled that of an ape, with big eyebrow ridges and a flattened nose.

Homo habilis (Handy Man) was taller, standing about 5 feet in height. This species had a larger brain than the man-apes that came before them; this is judged on the size of the fossilized skulls. *Homo erectus* (Upright Man) had an even bigger body and brain than Handy Man, being 5–6 feet tall, and weighing 90–160 pounds.

These people looked remarkably like modern humans, although the face shape was different. Upright Man's brain was midway in size between that of Handy Man and our own. The Neanderthal people evolved around 250,000 years ago. They probably died out about 30,000 years ago when the modern sub-species of *Homo sapiens,* such as the Cro-Magnon people, became dominant.

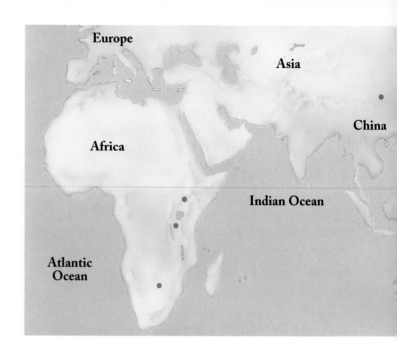

Where early humans lived
The map shows where fossilized remains of the earliest Homo *species have been found. The middle site in Africa is the famous Olduvai Gorge in Tanzania. Hominid remains have also been found in central China.*

Australopithecus afarensis

Homo habilis (Handy Man)

Homo erectus (Upright Man)

Where did the first humans live?

The only way scientists can work out where ancient humans lived is to plot where their fossils and tools have been discovered and can still be found under the ground today.

It seems certain that we began in Africa, since the remains of our early *Australopithecus* relatives have all been found there. There is even direct evidence that our ancestors were walking upright on two legs on the plains of what is now Tanzania millions of years ago. In 1976, scientists working with Dr. Mary Leakey in Tanzania found two long tracks of man-ape footprints made by an adult and a child nearly four million years ago.

Homo habilis (Handy Man) is known from fossils found in East Africa and southern Africa, and there is no clear evidence that these early humans ever reached Asia. *Homo erectus* (Upright Man) also seems to have evolved in the African continent, but these peoples certainly spread farther afield in search of food and shelter. Fossils, including the famous "Peking Man" and "Java Man," show that they reached Asia, and they probably also lived in southern Europe.

By 300,000 years ago, Upright Man had evolved into a new species. Most scientists believe that this was an early version of modern humans. An example of an early *Homo sapiens* fossil is the female skull found at Swanscombe, near London, England. The early fossils of our own species show that they lived in all the places where Upright Man had reached.

More modern versions of our species, including the famous Neanderthal people, are best known from Europe and the Middle East.

Homo sapiens
(Neanderthal)

Homo sapiens
(Cro-Magnon)

Homo sapiens

Evolution of humans
Landmarks in the evolution of modern people are shown in this series of drawings. On the far left is the tiny **Australopithecus afarensis,** *the species to which "Lucy" belonged (see page 25). The remaining examples are all members of the same group,* **Homo,** *which includes ourselves.* **Homo habilis** *is an early species from Africa that certainly used tools.*

Why are there different races of humans?

The differences in the many races of human beings, like the variations between the subspecies of any animal, arise because living things change through time. They may become altered or adapted to survive in their local surroundings. The main differences are easy to spot and have to do with body build, hair, skin type and color, and the shape of the face.

Humans are distributed more widely than any other backboned animal. We spread all over the surface of the globe, except for Antarctica, long before modern forms of transportation were invented, and by about 10,000 years ago lived on most parts of the other continents.

Among all sorts of animals, types that spread over wide areas nearly always end up with populations adapted for the geographical area in which they live. These are the different subspecies, or races, of humans. You can tell from the pictures of people here roughly where their ancestors came from.

The color of a person's skin is a good example of this kind of adaptation. Dark brown or black skin is useful in places with strong sunshine because it stops the harmful ultraviolet rays in the sunlight from causing skin cancers. In less bright light, this protection is not so important. Indeed, in low sunlight, dense skin coloration (brown or black) can stop the light from making vitamin D in the skin. So, as early humans migrated from Africa, northern peoples lost much of their skin pigment, which was a barrier to sunlight, and became paler skinned. In the tropics, there is enough light to form vitamin D even if the skin is dark.

People all over the world now travel much more than they used to, and marriages between people from different racial groups are more common. These changes are likely to make racial differences increasingly blurred.

Inuit (Eskimo) woman

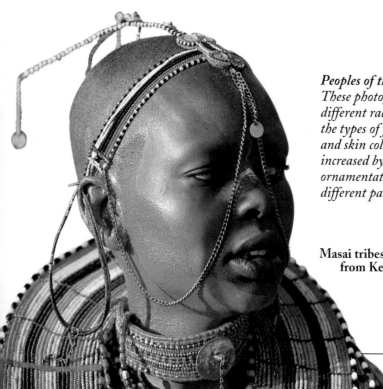

Masai tribeswoman from Kenya

Peoples of the world
These photographs of people typical of different racial groups show some of the types of facial characteristics, hair, and skin color that exist. Contrasts are increased by the clothing and body ornamentation used by people in different parts of the world.

European woman

How did humans spread all over the world?

Humans of different species have spread around the world at different times. In each case the fossil evidence suggests that the new type of human developed first in Africa, before spreading to other parts of the world. For this reason, Africa has been called the "cradle of humankind."

The first movements happened with the early human species *Homo erectus*, and later with *Homo sapiens* (see pages 26–27). The *Homo sapiens* group evolved in Africa about 1.8 million years ago and later spread to eastern Asia, what is now Indonesia, and probably Europe.

A similar pattern happened with the ancient form of *Homo sapiens*, which appears to have given rise to modern versions much like ourselves in Africa around 100,000 years ago. It is the descendants of this African-based new human type that eventually colonized the whole world and caused all the other human types that were there before, including the Neanderthals, to die out.

The evidence for this theory was based first of all on fossil bones of early humans and findings of artifacts such as stone tools and weapons. But more recently, it has become possible for scientists to compare the genes of living people in different parts of the world. Tiny variations in their genetic makeup reflect people's ancestry, and these studies so far suggest that Africa is probably the home of all modern humankind.

The spread of humans
The map shows the spread of Homo sapiens *from Africa during the last 100,000 years. By 35,000 to 40,000 years ago, modern humans had spread throughout Europe and Asia and into Australia. Possibly people came to the New World from Asia shortly afterward.*

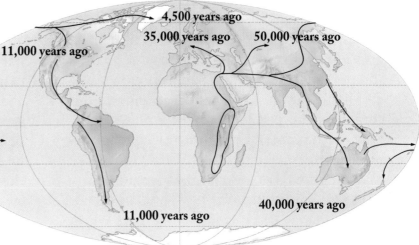

4,500 years ago

35,000 years ago

50,000 years ago

11,000 years ago

11,000 years ago

40,000 years ago

Peoples of the world
The Chinese girl and Aborigine shown here are typical of their races. But increasing numbers of people in the world cannot be identified so easily, because their parents come from different races. Such people inherit characteristics from both parents.

Chinese girl

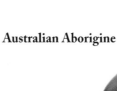

Australian Aborigine

When did people first become farmers?

The first three crop-growing areas of the world.

Farming started more than 10,000 years ago, and its discovery changed the lives of our ancestors. It began as the world moved out of the cold phase of the last Ice Age. Around this time, the glaciers and ice caps began to melt, the sea level rose, and areas which had once been either barren desert or tundra became grasslands or forests. In this gentler climate, a few groups of early humans began to grow plant crops for eating, and to keep animals to use as food.

For the 4 million years before that, man-apes and earlier species of humans were "hunter-gatherers." They had a very mixed diet based on the small and large animals they could catch and kill and all the edible plants they could gather in the forest. To live in this way meant being constantly on the move, following herds of animals, and seeking out trees and bushes in fruit.

The change to a farming way of life started only in a few fertile areas such as the Middle East. Other groups of people kept to the hunter-gatherer lifestyle, and some— such as the Pygmy people in Central African jungles and the peoples in the Amazon rain forest in South America—are still hunter-gatherers to this day.

Primitive methods of farming, using animals instead of machines, are still employed in developing parts of the world.

What were the first crops?

The earliest farmed crops seem to have been a number of grasses of the types we now call cereals. There is evidence from 13,000 years ago in Israel that people were harvesting wild grass seeds as a nutritious and easily stored food, using sharp-edged sickles.

From this way of life, people in different areas probably began to grow edible grasses deliberately. They sowed the first crop fields by scattering their seed on the ground near to their settlements. The harvest would have produced seeds to be stored for sowing the following year.

Preserved seeds from early settlements show that this type of farming became common around 10,000 years ago.

Wheat

Barley

A little later—about 7,000 years ago—people in the region of China farmed millet, rice, soybeans, and yams. Later still—5,000 years ago—in the warm, wet parts of Central America, maize (sweet corn), beans, and cotton were cultivated.

What were the first farmed animals?

Aurochs (wild cow)

Wild boar

Red jungle fowl

Mouflon (wild goat)

Przewalski's horse

They were animals not unlike those of today—sheep, goats, cows, and pigs. They were tamed from wild sheep and goats, wild cattle called aurochs, and from wild boar. In Southeast Asia, the red jungle fowl was also domesticated as the chicken and became an important source of meat and eggs. Goats, sheep, and cattle could feed on grass and tough plants that were no use as food for humans, while pigs could eat up any food scraps that were available.

When the hunter-gatherer peoples killed adult prey animals, they must have been able to take their young, too, at an age when they were easy to control. This may have been how the first animals became domesticated. These farmed animals would have been what some scientists call "living larders," a source of valuable meat, fat, and hides that could be used at any time they were needed, simply by killing an animal. Their milk would also have been a useful source of food. This must have made a great difference in periods of the year when food was otherwise scarce.

Sheep are thought to have been first kept as domestic animals in the Middle East 8,000 years ago, and pigs, cattle, and goats from about 7,000 years ago.

The Living World

Humans divide the world into countries. Nature divides it into areas according to how hot, cold, wet, or dry they are and what plants grow there. These natural areas are the Earth's habitats and include tundra, forest, woodland, and desert. Every habitat is different because of its climate and the sorts of plants that can live there. It may be hot and dry with few plants, as in the desert, or hot and very wet with an enormous variety of plants, as in the tropical rain forest. In addition to the land habitats, there are the freshwater and ocean kingdoms with their own special inhabitants.

Within these habitats are an extraordinary variety of creatures, each with its own way of life. In this chapter, we discover the answers to some of Nature's puzzles, such as why tigers have stripes, why trees lose their leaves, and why hummingbirds hum.

Antarctic birds
King penguins (left) are
perfectly adapted to survive in
the harshest of all habitats—the
Antarctic. Their thick feathers
protect them from cold and wet.

Why does the world have so many different animals and plants?

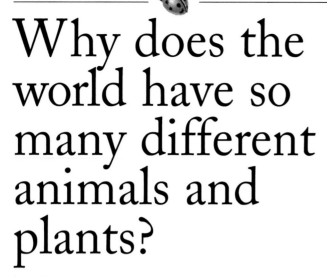

There are certainly more than a million species of living things in the world—plants, animals, and microbes. Some biologists have suggested that ultimately there could be as many as 30 million species waiting to be discovered.

The number is high because each kind of plant or animal—that is, each species—has its own special lifestyle or "niche." (A niche is a total description of where and how an organism lives.) Because there are millions of possible niches in the world, there are similar numbers of species.

The reason the number of niches is so great is because climate, landscape, and opportunities vary so much from place to place. In addition, the plants that flourish in different regions produce a whole new set of niche possibilities. The plants provide food, shelter, and nest sites for many of the animals and microbes that live in the same area.

In a similar way, every animal is a "home" to other creatures that live in or on it. Some of these are damaging, such as parasites and germs; others

are helpful partners that, for example, feed on the parasites. All these "hangers-on" are using a new set of niches based on the bodies of living things. Each species of animal has its own special set of parasitic worms and microbes that live on that species alone. These sorts of lifestyles multiply the number of niches available even further.

With all these possibilities for different ways of life, there seems to be almost no limit to the variety of living things that Earth can support.

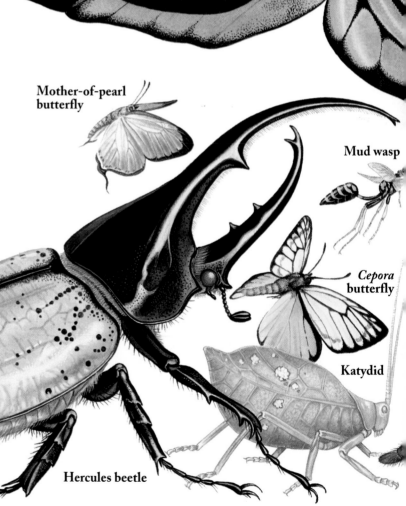

German cockroach

Mother-of-pearl butterfly

Mud wasp

Cepora **butterfly**

Katydid

Lamprima **beetle**

Hercules beetle

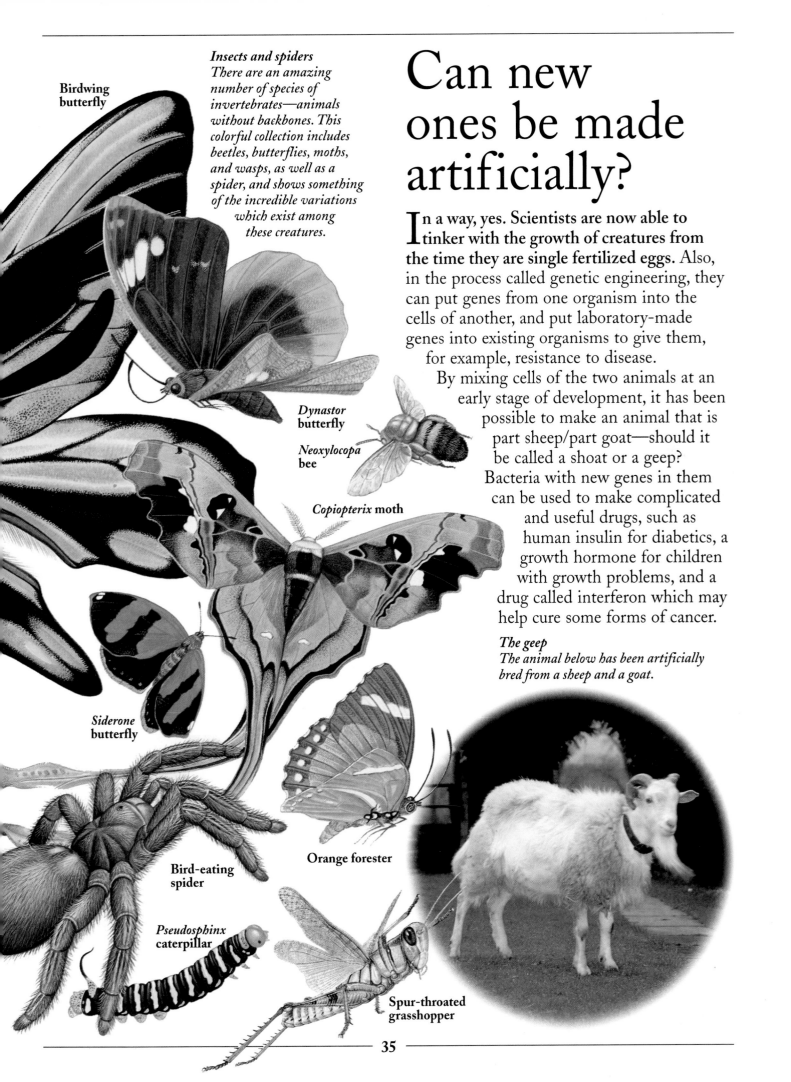

Birdwing butterfly

Insects and spiders
There are an amazing number of species of invertebrates—animals without backbones. This colorful collection includes beetles, butterflies, moths, and wasps, as well as a spider, and shows something of the incredible variations which exist among these creatures.

Dynastor **butterfly**

Neoxylocopa **bee**

Copiopterix **moth**

Siderone butterfly

Bird-eating spider

Pseudosphinx **caterpillar**

Orange forester

Spur-throated grasshopper

Can new ones be made artificially?

In a way, yes. Scientists are now able to tinker with the growth of creatures from the time they are single fertilized eggs. Also, in the process called genetic engineering, they can put genes from one organism into the cells of another, and put laboratory-made genes into existing organisms to give them, for example, resistance to disease.

By mixing cells of the two animals at an early stage of development, it has been possible to make an animal that is part sheep/part goat—should it be called a shoat or a geep? Bacteria with new genes in them can be used to make complicated and useful drugs, such as human insulin for diabetics, a growth hormone for children with growth problems, and a drug called interferon which may help cure some forms of cancer.

The geep
The animal below has been artificially bred from a sheep and a goat.

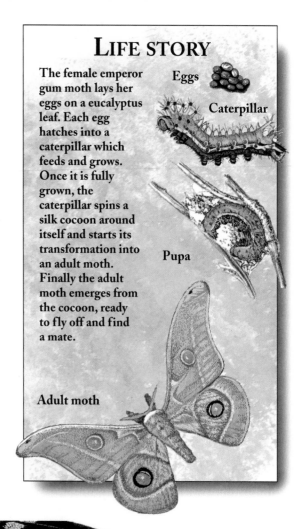

Why do caterpillars look so different from butterflies?

Although all true caterpillars turn into butterflies or moths, they do not look like them because they live in a completely different way. Caterpillars crawl around munching leaves with their strong jaws, while butterflies fly from flower to flower sucking nectar through their strawlike mouthparts.

The butterfly starts life as an egg laid on a plant. One to three weeks after the egg is laid, a tiny caterpillar as fine as a hair emerges from its egg and starts to feed. It eats almost constantly and grows so fast that it needs to shed its skin several times as its body gets bigger. At this stage of its life, all the caterpillar's energies are directed toward growth. When it is fully grown, the caterpillar stops feeding and starts the pupa, or chrysalis, stage. To prepare for this, it burrows into the soil or simply finds a safe spot and hangs from a little pad it spins from silk; some moth caterpillars spin complete cocoons of silk around themselves in which to pupate.

Once settled, the caterpillar begins its spectacular transformation into an adult butterfly, and antennae, legs, and wings start to form inside the pupal case. When all is ready, the pupa splits, and the butterfly wriggles out to start a new stage in its life. Now all the insect's energies are directed toward finding a mate and producing eggs.

Monarch butterfly

Monarch butterfly
The monarch takes about four weeks to change from egg to striped caterpillar to butterfly. The caterpillar feeds on poisonous milkweed plants. This poison does not harm the caterpillar, but it does make it taste horrible to birds.

Monarch caterpillar

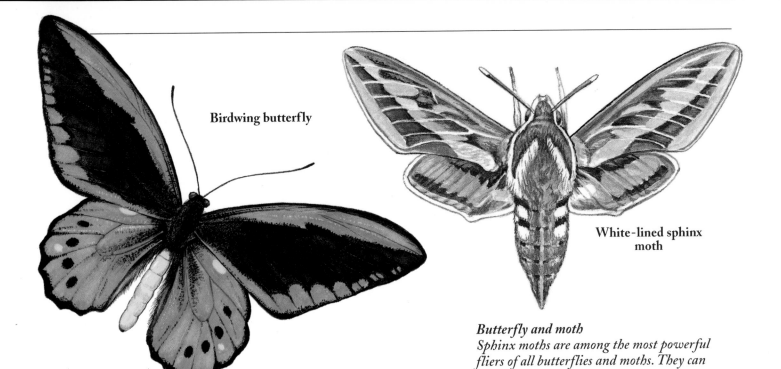

Birdwing butterfly

White-lined sphinx moth

Butterfly and moth
Sphinx moths are among the most powerful fliers of all butterflies and moths. They can even hover like hummingbirds in front of flowers as they feed. Birdwings, which live in the jungles of New Guinea, are the largest of all butterflies. They measure up to 11 inches across with their wings fully spread.

What is the difference between butterflies and moths?

Butterflies and moths belong to the same group of insects, the Lepidoptera, and there are no hard and fast differences between them. There are about 150,000 species of butterflies and moths. They range from tiny creatures a fraction of an inch long to the largest tropical butterflies which measure up to 11 inches across.

Typically, we think of butterflies as brightly colored insects that fly in the daytime and moths as drab nighttime creatures. But many moths, such as hawkmoths, are very beautiful, and some even fly by day. And there are plenty of dull brown butterflies. Most butterflies and moths are very similar in structure. All have two pairs of wings covered in tiny scales. They have large eyes, and most have sucking mouthparts made up of a long, coiled tube called a proboscis, through which the butterfly or moth sucks liquid foods.

The lives of butterflies and moths are similar, too. All go through caterpillar and pupa stages before becoming winged adults. As adults, both feed mostly on flower nectar or other liquids such as tree sap.

CATERPILLARS

The caterpillar of the elephant hawkmoth has markings that look like big eyes on the sides of its body. If in danger of attack by a hungry bird or some other hunter, the caterpillar pulls in its head, swelling up the front of the body and making the "eye spots" even bigger. These alarming big "eyes" may frighten off the bird which thinks them part of a much bigger creature! The cracker butterfly caterpillar has long poisonous spines. These can be used to inject poison into enemies and cause an irritating rash if touched.

Eye spots

Cracker butterfly caterpillar

Elephant hawkmoth caterpillar

Songbirds

The eastern meadowlark sings its song—a series of clear, whistling notes—from a fencepost or other perch or in a fluttering song-flight. The zitting cisticola, too, sings during a song-flight. It may go up to 100 feet or more above the ground and circle over a wide area of territory. The European robin sings its warbling, trilling song for much of the year.

Zitting cisticola

Eastern meadowlark

European robin

Do all birds sing?

All birds have voices, but only some birds sing. Gulls, grouse, and woodpeckers, for example, make a variety of calls which tell their neighbors that there is danger approaching or a good food supply to be found nearby. These calls do not have to be learned—they form an inborn part of the birds' behavior.

A large group of birds, called the passerines, or perching birds, can make a much wider range of sounds. These are the songs so admired by humans. Mainly performed by the males, the songs include information calls, but these are combined with elaborate trilling sounds which may vary from one individual to another. The basic pattern of the song is the same for all birds within any one species.

Most birds sing only when they are setting up territory and looking for a mate. Once these tasks are completed, their singing almost ceases. Although the basis of a song is inherited, a bird only learns how to sing perfectly after it has heard other birds of the same species singing.

Beautiful songs do not necessarily go with beautiful birds. Some of the most elaborate songs are produced by drab birds such as the sedge warblers.

NOISY BIRD

The laughing kookaburra, a giant kingfisher which lives in Australia, is one of the noisiest of all birds. It does not sing, but has a very loud and unusual call which sounds like wild human laughter. The kookaburra also makes some softer chuckling calls.

Laughing kookaburra

How do parrots talk?

Exactly why parrots can change their calls to make them sound like words is still not understood. Their ability may possibly be linked with the fact that almost all are highly social, forest-living birds which keep in contact with other members of their species by frequent calls.

A young parrot kept on its own in captivity learns the sounds it hears around it and quickly realizes that repeating these sounds brings attention and companionship. These are perhaps a substitute for its normal social life. The parrots' talents have caused it problems, however. Those with a particularly high reputation for mimicking human speech, such as the yellow-crowned amazon, have been targeted by illegal trappers for the pet trade and are now very rare.

In the wild, parrots make a variety of calls, such as squawks, screeches, and whistles. Scarlet macaws, for example, make loud metallic screeching calls as they fly in search of food, but they always feed in silence. Smaller parrots make softer twittering sounds.

Although they are such good mimics in captivity, parrots do not imitate other sounds in the wild. Many other species, such as mynah birds and lyre birds, do mimic the sounds they hear in their everyday lives.

Jungle bird
The scarlet macaw lives in rain forests in South America. In flocks of up to 20 birds, macaws feed on leaves, fruits, and seeds high in the trees.

Scarlet macaw

Gray parrot

Crimson rosella

Parakeet

Swift parrot

PARROTS

The parrot family includes more than 300 different species such as macaws, parakeets, rosellas, and lorikeets. Most are brightly colored, tree-living birds with short, hooked beaks. The beak is very strong and can be used almost like a third limb when climbing in trees. The powerful beak also helps parrots to crack hard foods such as nuts and seeds.

Do humming-birds hum?

Bee hummingbird

Smallest bird
The bee hummingbird is the smallest bird in the world. The male is 2 inches long including its beak and tail, but its body is only ¹/₂ inch long. It weighs only ¹/₁₅ ounce. The female is slightly larger.

Hummingbirds do hum, but the low-pitched buzzing sound is made by their wings, not their voices. When they fly, hummingbirds beat their wings constantly and very fast—in some species, more than 50 times a second. These continuous wing movements make an uninterrupted hum which gives the birds their name.

The flight of hummingbirds is much more energetic than that of other birds. Their wings are not large for their body size, but their flying muscles are relatively huge. These powerful muscles provide the energy for hummingbirds to beat their wings so fast that they can hover in the air, and even fly backward as well as forward.

Sword-billed hummingbird

Male

Female

Giant hummingbird

Frilled coquette

Collared inca

Why do hummingbirds hover?

A hummingbird hovers as it feeds. Nectar from flowers is its main food, and since there is rarely a convenient perch in front of the flower, the bird needs to hold itself in midair while it feeds. It can only do this by the exhausting process of hovering. When a hummingbird hovers, its body is held vertically, and its wings move backward and forward, not up and down as in normal flight.

Male

Raggiana bird of paradise

Female

A group display
Raggiana birds of paradise display in groups. When females appear, the males hop around, flapping their wings and calling. Each tries to outdo the others in showing off his glorious plumage.

How did birds of paradise get their name?

The birds were named by European travelers who began to explore and trade with the East in the nineteenth century. Among items brought back from New Guinea were bird skins, with full plumage but usually with the feet and legs cut off by the local hunters. Some of these birds were so beautiful that people called them birds of paradise. They thought them too exquisite to be earthly birds and imagined them constantly airborne in paradise with no need of feet for perching.

Male birds of paradise are brilliantly colored, often with long, filmy or iridescent plumes. They use their beautiful plumage in courtship displays performed to attract the much duller colored females. Male birds may even hang upside down from branches to show off their fine feathers.

Male

Upside down display
The blue bird of paradise hangs upside down in his efforts to attract females. His long tail streamers form a graceful arc over his magnificent plumage.

Blue bird of paradise

Female

Wandering albatross
This great sea bird spends much of the year circling the southern oceans, searching for food. It nests on oceanic islands.

Breeding areas

Short-tailed shearwater
This shearwater breeds on islands near Tasmania in early spring. It then spends the next seven months flying around the Pacific Ocean, feeding on sea creatures that it plucks from the ocean surface.

Breeding area

Why do birds migrate?

Breeding and feeding are the two main reasons that birds migrate. For most species, migration is a twice-yearly journey made from a summer breeding area, in which food is plentiful, to a warmer wintering place and back again. By migrating between two parts of the world in this way, birds can get the best of both.

Barn swallows, for example, live in most of North America and Europe in summer, feeding on the swarms of flying insects. But the winter weather is so severe over most of their breeding range that the birds move south to areas where they can continue to find food. Barn swallows that breed in warmer parts of the range such as southern Spain do not migrate, but stay in their breeding range all year round.

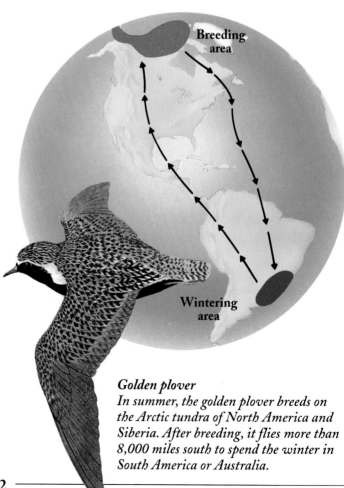

Breeding area

Wintering area

Golden plover
In summer, the golden plover breeds on the Arctic tundra of North America and Siberia. After breeding, it flies more than 8,000 miles south to spend the winter in South America or Australia.

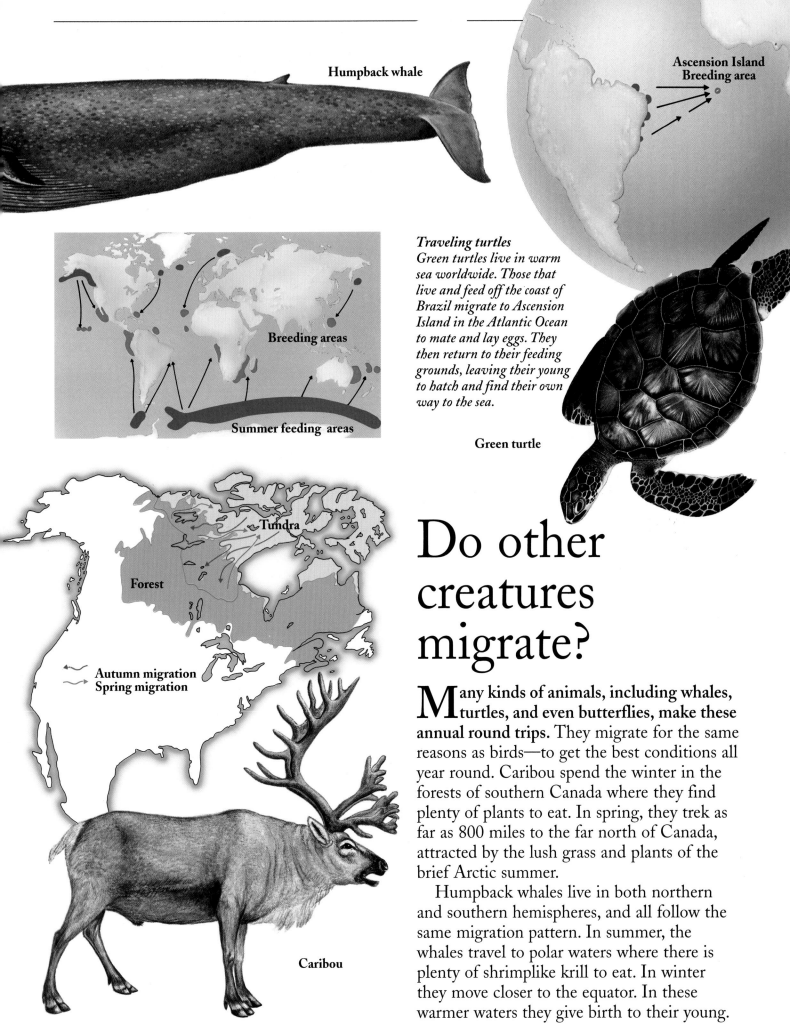

Humpback whale

Ascension Island Breeding area

Breeding areas

Summer feeding areas

Traveling turtles
Green turtles live in warm
sea worldwide. Those that
live and feed off the coast of
Brazil migrate to Ascension
Island in the Atlantic Ocean
to mate and lay eggs. They
then return to their feeding
grounds, leaving their young
to hatch and find their own
way to the sea.

Green turtle

Tundra

Forest

Autumn migration
Spring migration

Caribou

Do other creatures migrate?

Many kinds of animals, including whales, turtles, and even butterflies, make these annual round trips. They migrate for the same reasons as birds—to get the best conditions all year round. Caribou spend the winter in the forests of southern Canada where they find plenty of plants to eat. In spring, they trek as far as 800 miles to the far north of Canada, attracted by the lush grass and plants of the brief Arctic summer.

Humpback whales live in both northern and southern hemispheres, and all follow the same migration pattern. In summer, the whales travel to polar waters where there is plenty of shrimplike krill to eat. In winter they move closer to the equator. In these warmer waters they give birth to their young.

Why are tropical fish such bright colors?

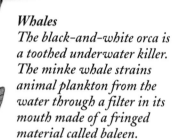

Mandarinfish

Biologists do not fully understand why fish in tropical waters, particularly those in shallow seas and coral reefs, display such a dazzling mixture of patterns and colors. But it is a fact that, as you go from cold seas to warmer seas and toward the tropical seas near the Equator, the fish become more and more brightly colored. The colors are caused by a patchwork of color-filled cells in the fish skin. Each one of these is filled with brilliant natural pigments.

The gaudy hues of some tropical fish act as signals to other animals. The lion fish, for example, has a striped body and fins. The colors act as a warning that this fish has extremely poisonous spines on its fins. Other fish species learn to associate the danger of this fish with its patterns. The clown anemonefish is also brightly colored, probably as a warning to predators, but it is not poisonous itself. Instead, it lives among the stinging tentacles of sea anemones, to which it is immune. Its bold stripes remind other fish that an attack will involve braving the anemone's tentacles.

Markings may also protect fish from enemies. The dark "eye spots" near the tail of the butterflyfish and the stripes of the angelfish may distract predators from the vulnerable head area.

The bright shades of many other tropical fish help males and females of the same species to recognize each other as mates. There is a greater need for this in the tropics than in cooler waters, since these warm seas have a far greater number of different sorts of fish in them. More species means more possible confusion about mating partners.

Whales
The black-and-white orca is a toothed underwater killer. The minke whale strains animal plankton from the water through a filter in its mouth made of a fringed material called baleen.

Sweetlip emperor

Clown anemonefish

Lion fish

Imperial angelfish

Copperband butterflyfish

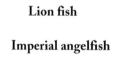

Queen angelfish

44

Minke whale

Orca

How do whales breathe?

A whale is a mammal, and like all other mammals, it breathes in air through its nostrils. The air passes into its lungs where oxygen is removed. Unlike other mammals, its nostrils, called a blowhole, are on the top of its head. As a whale dives, the blowhole is tightly closed against the water and remains shut all the time the whale is submerged.

When the whale finally surfaces and breathes out, the moisture in its warm breath condenses in the cooler air and makes a cloud of water vapor, just as our breath does on a cold day. A huge new breath is then taken in through the same hole before the whale dives again. When they dive, whales carry little oxygen in their lungs, but large amounts in their blood, muscles, and other body tissues. The deeper they dive, the longer they must spend breathing at the surface between dives.

Contrary to the impression given in many old drawings of whales, they do not spout water from their blowholes.

Deep diver
Only its tail fins are visible as a humpback whale dives beneath the surface in search of food. Humpbacks usually stay under for between 3 and 28 minutes.

How deep do whales dive?

One sperm whale was found entangled with a submarine cable at a depth of about half a mile, and other species may dive as deep. Sperm whales rarely dive for less than 30 minutes and may stay under 90 minutes or more, hunting for giant squid to eat. After a deep dive, the whale stays at the surface for ten minutes or more, breathing deeply.

As a rule, filter-feeders, such as the minke whale, find their food near the surface of the sea and rarely remain submerged for more than 15 minutes. Small whales, such as dolphins, seldom go below 500 feet.

Where do you find coral reefs?

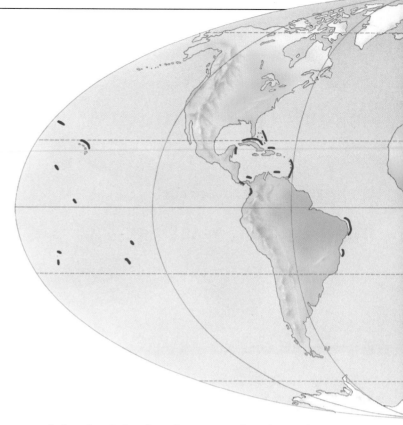

Coral reefs are found in the warm seas of the tropics. As the map shows, reefs are common around the tropical islands of the Indian and Pacific oceans. In the same region is the famous Great Barrier Reef in Australia. There are coral reefs down parts of the East African coast, and they also stretch as far north as the northern end of the Red Sea. There are also beautiful reefs in the Caribbean and on some South American coasts.

Coral reefs are massive underwater structures made of rocky material similar to that which makes up marble or chalk. But this "rock" is built by living sea creatures—the coral animals known as polyps. They are closely related to sea anemones and jellyfish.

There are three different types of reefs found in the warm waters of the world. They are called barrier reefs, fringing reefs, and atolls. The Great Barrier Reef stretches for 1,250 miles down the eastern coastline of Australia. It is a huge reef that runs parallel to the shoreline, but is separated from it by a considerable distance of 60 miles or so. Fringing reefs are usually much closer to the shore. Most are found around islands in the middle of the ocean.

Atolls are circular reefs with no island in the middle. Instead, in the center of the reef ring is a shallow lagoon. Scientists think that an atoll is formed when an island with a fringing reef slowly sinks (or the ocean level rises). The island disappears beneath the waves because it cannot grow upward. The reef, though, is living and can grow up to keep pace with the changing sea level.

Coral reefs are found only in warm saltwater and in shallow seas, which receive plenty of sunlight. They need clean seawater, unmuddied by silt or sediment. This is why reefs are never found at great sea depths, nor are they found on the coasts near river mouths and estuaries where there is too much silt.

There is a greater variety of sea creatures living on a coral reef than in any other marine habitat. In addition to a host of colorful tropical fish, there are many other creatures, including starfish, sea anemones, sponges, sea urchins, and many kinds of crustaceans.

Coral polyps

Coral animals
At the base of the coral animal is the rocky skeleton of the polyp. It is these skeletons that build the reef. Growing up from it into the water is the polyp itself. Thousands of these live joined to one another in large colonies. The polyp uses its tentacles to catch small animals which are then passed down to its mouth. The tentacles are lined with tiny stinging cells.

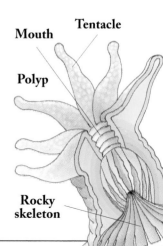

Mouth

Tentacle

Polyp

Rocky skeleton

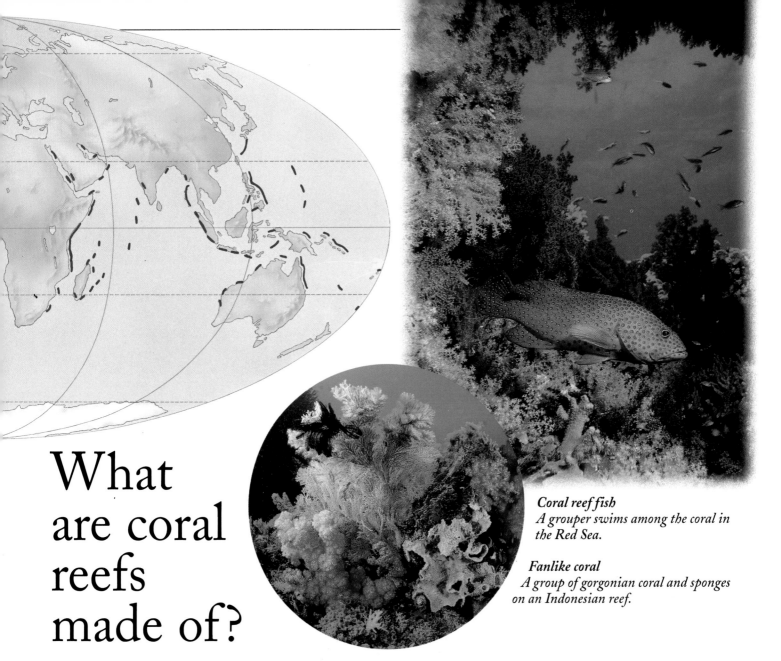

Coral reef fish
A grouper swims among the coral in the Red Sea.

Fanlike coral
A group of gorgonian coral and sponges on an Indonesian reef.

What are coral reefs made of?

Coral reefs are made from the rocky skeletons of coral animals. Inside each skeleton is a creature rather like a minute sea anemone called a polyp. The polyps live together in huge colonies of many thousands of individuals. This mass of animals, unlike ordinary sea anemones, makes a hard rocky base under itself, and around each of the polyps, as protection.

Vast numbers of these coral masses make up a reef, just under the water surface. The polyps themselves poke out from tiny holes or pores in the rocky structure to feed with their tentacles. They do this by stinging their prey, then curling their tentacles over to capture it. Some of the rock of coral reefs is made by special types of seaweeds that lay down rock material like the polyps do.

Most coral reefs are found in shallow water because of a strange partnership that exists in all corals. Tiny plants (algae) live inside the coral animals. Without these plants, the corals cannot make their rocky skeletons. Only in shallow waters is there enough light for the algae to live and to trap sunlight energy in the process of photosynthesis.

There are many species of coral in a wide variety of shapes, sizes, and colors. Some are smooth, rounded structures, such as brain corals. Others, such as the stag corals, stick up into the water in complicated branching shapes resembling a deer's antlers. Gorgonian corals spread like great fans in the water. Not all types of corals build reefs, however. Soft corals do not have hard rocky skeletons, so they do not add to the structure of the reef.

What is an insect?

Insects are invertebrate animals—they do not have backbones. In terms of numbers, they are the most successful creatures on Earth. At least a million species are known, and there are thought to be millions more yet to be discovered. Most are small—the most common size is about $\frac{1}{4}$ inch long—but they have colonized every type of habitat and eat every imaginable type of food. Because they are so small, insects can use a vast range of microhabitats that would be unsuitable for larger creatures—hundreds of different species can live in one tree, for example.

An insect's body has three parts—head, thorax, and abdomen. The head carries the eyes, mouthparts, and a pair of sensory antennae, which the insect uses to find out about its surroundings. The mouthparts may be designed for chewing food or for sucking or lapping up liquids. The thorax bears the insect's three pairs of legs and, usually, two pairs of wings. The abdomen contains the reproductive organs and most of the digestive system. A tough waterproof layer outside the body called an exoskeleton protects the insect.

What is a fish?

More than half of the 43,500 or so known species of vertebrates (animals with backbones) are fish. They are divided into three main groups. The first most primitive group includes the lampreys and hagfishes. These fish have no jaws, only suckerlike mouths. The second group includes all the sharks and rays. These are known as cartilaginous fish because their skeletons are made of a gristly substance called cartilage, not bone. The third, and largest, group contains about 20,000 different species of bony fish. As their name suggests, these fish have skeletons made of bone.

All fish spend their lives in water and have gills which allow them to take oxygen from water, not air. Instead of legs, they have fins and a tail to help them move through water. They feed in a wide variety of ways. Some eat aquatic plants. Others catch tiny animals or strain them from the water through filterlike structures attached to the gills. Many are active, fast-moving hunters, with sharp teeth, while others, such as flatfish and angler fish, lie hidden on the seabed and wait for prey to come close enough to catch.

What is an amphibian?

Amphibians were the first vertebrate animals to live on land. They evolved from fish, and modern amphibians still spend part of their lives in water. Most lay their eggs in water. These hatch into aquatic tadpoles which have fins and gills. As the tadpoles grow, they develop legs and lungs to enable them to come out onto land.

Amphibians are cold-blooded—they cannot control their own body temperature and must gain heat by basking in the sun. Their skin is smooth, not scaly, but must be kept moist. Adult amphibians do have lungs, but they also breathe through their skin. There are only about 2,000 species of amphibians, divided into two main groups—salamanders and newts, and frogs and toads.

What is a reptile?

Reptiles are vertebrate animals, but they are not warm-blooded. They cannot control their own body temperature and rely on the sun for warmth. As a result, most of the approximately 6,000 different species of reptiles live in warm climates.

The first reptiles evolved from amphibians about 300 million years ago. In the form of dinosaurs, reptiles were the dominant animals on Earth for more than 130 million years (see pages 16–23).

Reptile eggs show great advances over amphibian eggs. They do not have to be laid in water and have tough shells to protect them from damage and water loss in the soil, where they are normally laid. Each egg contains yolky food stores to nourish the baby reptile while it develops directly into a miniature adult.

Four groups of reptiles survive today. Turtles and tortoises have short broad bodies, enclosed by a bony box into which head, tail, and limbs can be pulled for protection. The second group, crocodiles and alligators, are the only remaining representatives of the archosaurian reptile group to which dinosaurs belonged. All crocodiles are hunters and are the largest living reptiles. The third group includes the lizards and snakes. All snakes and most lizards are predators. The last group, the Rhynchocephalians, has only one living representative, the lizardlike tuatara, which lives in New Zealand.

What is a bird?

A bird, too, is a vertebrate, an animal with a backbone. There are more than 9,000 species of birds known. The vast majority can fly, and their structure reflects this. A typical bird has a strong but light body, two legs, and a pair of wings. All birds are covered with feathers which keep them warm and streamline the body. For many birds, the color of their feathers helps them to attract mates or hide from enemies. It is the large strong feathers on the wings and tail that mean birds can fly.

All birds lay eggs. It would be impossible for birds to carry their growing young inside their bodies as mammals do because they would become too heavy to fly. The embryo is protected inside a hard-shelled egg which is usually kept warm by the parents, normally in a nest. The incubation period—the time it takes for the fertilized egg to grow into a chick and break out of the shell—varies from about 10 days in small birds, such as warblers and wrens, to as many as 84 days in larger species such as albatrosses.

What is a mammal?

A mammal is a warm-blooded, hairy vertebrate (an animal with a backbone). Mammals evolved from reptiles and became the dominant land animals after the extinction of the dinosaurs 65 million years ago. There are about 4,000 different species known. Most mammals live on land, but some, including whales, dolphins, and seals, are adapted to live in water. Mammals in the form of bats have even taken to the air.

Some mammals, such as elephants and cattle, feed only on plant material. Others, including bats, anteaters, and shrews, catch and eat insects and other small creatures. Cats, dogs, and other carnivores kill other vertebrate animals, while a few such as rats eat almost anything they can find.

Typically, a female mammal keeps her developing baby inside her body. Some mammals, including rabbits, rodents, and many carnivores such as cats and dogs, are born naked, blind, and helpless, and are completely dependent on their parents. Others, such as cattle and deer, are born as small but fully formed versions of the adult and are able to move around shortly after birth. Female mammals feed their young on milk from their mammary glands.

Some mammals are a little different, however. Monotremes—the platypus and spiny anteaters—lay eggs like their reptilian ancestors. Kangaroos and other marsupials give birth to young which are very poorly developed. The young marsupial then finishes its development in a pouch on the mother's body (see page 60).

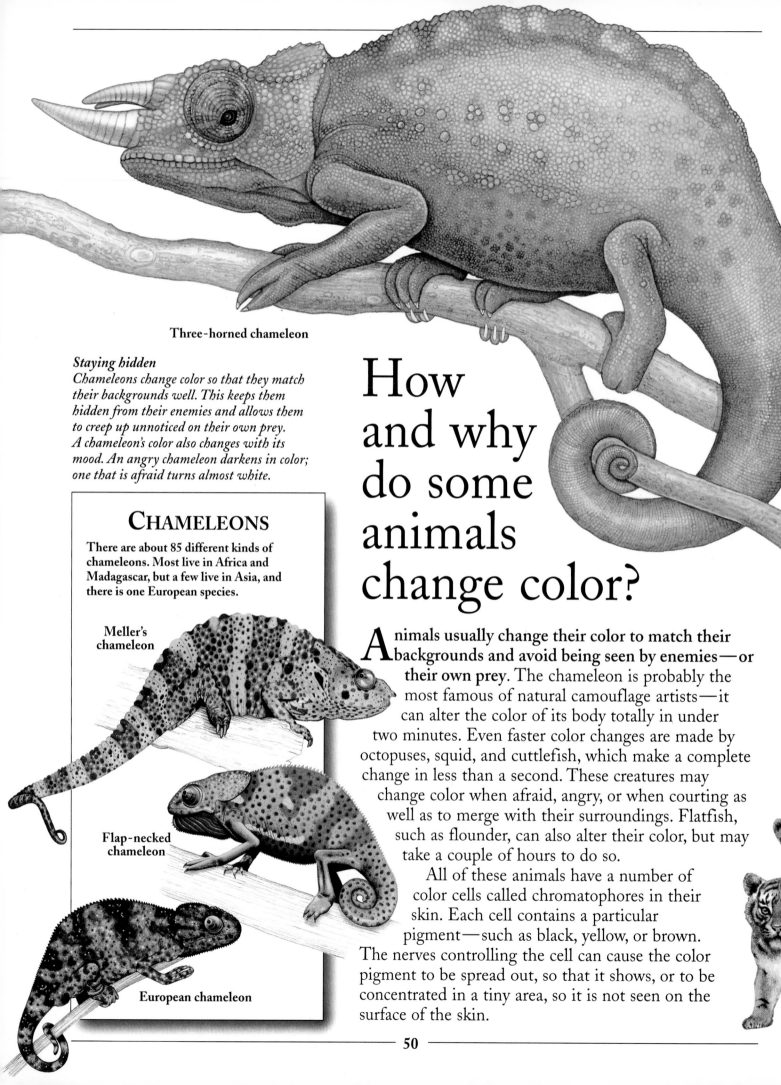

Three-horned chameleon

Staying hidden
Chameleons change color so that they match their backgrounds well. This keeps them hidden from their enemies and allows them to creep up unnoticed on their own prey. A chameleon's color also changes with its mood. An angry chameleon darkens in color; one that is afraid turns almost white.

CHAMELEONS

There are about 85 different kinds of chameleons. Most live in Africa and Madagascar, but a few live in Asia, and there is one European species.

Meller's chameleon

Flap-necked chameleon

European chameleon

How and why do some animals change color?

Animals usually change their color to match their backgrounds and avoid being seen by enemies—or their own prey. The chameleon is probably the most famous of natural camouflage artists—it can alter the color of its body totally in under two minutes. Even faster color changes are made by octopuses, squid, and cuttlefish, which make a complete change in less than a second. These creatures may change color when afraid, angry, or when courting as well as to merge with their surroundings. Flatfish, such as flounder, can also alter their color, but may take a couple of hours to do so.

All of these animals have a number of color cells called chromatophores in their skin. Each cell contains a particular pigment—such as black, yellow, or brown. The nerves controlling the cell can cause the color pigment to be spread out, so that it shows, or to be concentrated in a tiny area, so it is not seen on the surface of the skin.

Why do tigers have stripes?

The stripes which make a tiger so obvious in the zoo help it merge with its surroundings in its natural home of forest and scrubland. As well as blending with the lines of tall grasses in which the tiger hides, the stripes break up the body outline and make it harder for prey to see. Such camouflage is vital for the tiger's hunting success. The tiger is only able to approach within striking distance of its prey because it is so hard to see. Tigers live in Asia. Those that live in snowy Siberia in the far north have much paler fur than the tigers of the jungles farther south.

The zebra is the only other large animal with bold vertical stripes. It has been suggested that the zebra's stripes, too, are for camouflage but this is not so; in their open savanna home, zebras are very obvious. Recent studies show that zebras are extremely social animals and that their stripes, which are different in every animal, help them recognize their own family group and neighbors.

STRIPED COATS

Some forest animals, such as okapi and young wild pigs and tapirs, are camouflaged by horizontal lines on their bodies. On a tiger, such a pattern would not work. While the irregular vertical stripes merge with the lines of the tall grasses among which the tiger hides, horizontal stripes would cut across them, accentuating the outline of the body and making the animal more obvious to its prey.

Real tiger

Imaginary tiger

Forest cat
The jaguar's spotted coat helps to conceal it in the dappled light of the South American rain forest where it lives.

Tiger family
A mother tiger guards her young cubs, keeping them hidden in the long grass.

Leaf insect

Thorn bugs

Why is it so difficult to see animals in the wild?

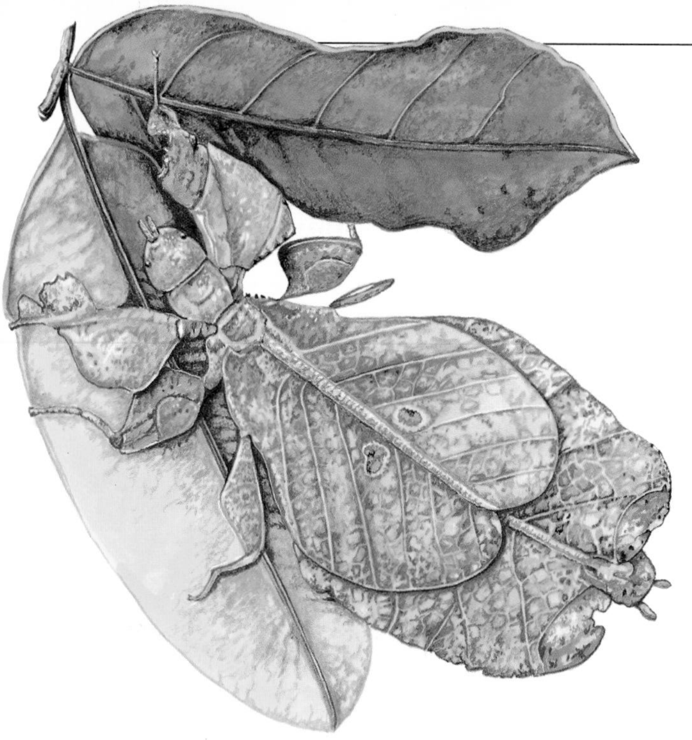

Good camouflage can make animals hard to spot in the wild. The color of an animal's coat or feathers often matches the surroundings in which it is normally found. Desert creatures tend to be sandy colored; arctic creatures, such as polar bears, are white. Most living places are more varied than these two, however, and many animals are striped and blotched with different shades. This keeps them hidden against the varied colors of most natural habitats.

Some creatures have spines and rough outlines which help to disguise them against the unevenness of their normal habitat. Animals which rely on camouflage for protection must keep still for long periods—any movement gives them away.

Many insects are colored to look like the plants they rest on. Brightly colored flower mantids, for example, are almost impossible to see as they lie in wait for prey among flowers. Crab spiders,

Common marbled carpet moth

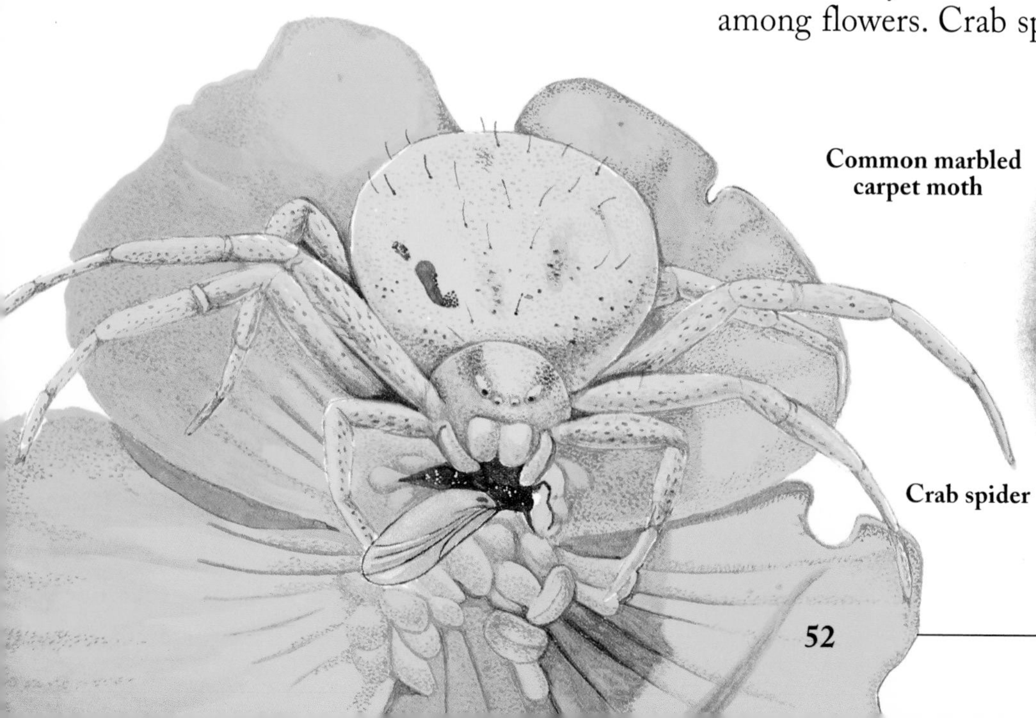

Crab spider

too, are colorful and sit on flowers that more or less match their body color as they watch for prey. The mottled colors of the common marbled carpet moth make it almost invisible on tree bark.

Leaf insects are not only green to match their leafy surroundings but also have wings shaped like leaves, with veinlike markings. Some even have irregular edges to their wings, making them look as if they have been nibbled by caterpillars. The strange-shaped bodies of thorn bugs look like thorns on a branch and probably also make the bugs more difficult for birds and other animals to swallow.

COLOR CHANGE

The fur of weasels which live in areas of heavy winter snowfall is brown in summer, but it changes to white in winter to match the changes in its surroundings.

Weasel

Summer coat   **Winter coat**

Matamata turtle

Stonefish
The warty skin of this fish is disguised with bits of coral, mud, and algae. If threatened, it raises the poisonous spines on its back. This fish lives on the seabed and half-buries itself in the sand, where it is very hard to see. Any small fish that swims by is quickly swallowed.

Jungle turtle
The matamata turtle lives in the murky waters of the Amazon where, with its ragged outline, it blends with dead leaves, wood, and other debris.

Flower mantis

Polar bear

53

Cheetah
70 miles an hour

Cheetah
Long, powerful legs help the
cheetah to run fast, but it is also
assisted by the great flexibility of
its back, which acts like a spring
to propel it on its way. The
cheetah runs fast to catch other
fast-running animals to eat,
such as gazelles.

Which animal is the fastest runner?

Of all animals, the fastest runner is the cheetah. It has been timed covering a distance of 700 yards in 40 seconds—equivalent to a speed of 70 miles an hour.

Many such animal speed records are calculated from speeds timed over short distances—animals do not usually run fast for long periods of time. Cheetahs, for example, do not normally run as far as a mile and usually start their high-speed attacks at a distance of less than 300 feet from their prey. The cheetah can reach its maximum speed in only 3 seconds but, like a human sprinter, cannot maintain this effort for long.

The North American pronghorn antelope cannot run as fast as the cheetah, but it can keep its speed for longer as it runs to escape from wolves and other enemies. It has been timed alongside a car traveling at 53 miles an hour and did not appear tired after nearly 4 miles. Over a short distance, the pronghorn has been clocked at 60 miles an hour. Even a young pronghorn can outrun a horse.

Racerunner lizard
18 miles an hour

Sailfish
68 miles an hour

How fast can other creatures move?

Ostrich
The world's largest bird, the ostrich is too big to fly, but is a high-speed runner. It is important for the ostrich to be able to move quickly so that it can escape enemies such as lions.

The peregrine falcon diving steeply on its prey is said to move at more than 100 miles an hour. The spinetail swift is believed to fly at about 70 miles an hour normally and to reach 100 miles an hour in display and courtship flights. Another fast flyer is the merganser duck, which flies at 40 miles an hour.

Among insects, a large dragonfly can probably reach 35 miles an hour in still air, but with a following wind flies at greater speed.

On land the lion, like the cheetah, can attain short bursts of speed. It can sprint at up to 40 miles an hour to seize its prey. An ostrich, the fastest-running bird, can move even faster. It runs at least 45 miles an hour and can possibly reach 60 miles an hour for short bursts.

Reptiles are well behind. The fastest lizard on record is the racerunner, timed for a minute at 18 miles an hour, while the fastest snake can only manage about 7 miles an hour.

Because water is 800 times more dense than air, it is difficult for animals to swim at high speeds. All the fast-swimming fish have streamlined bodies to cut through the water with minimum resistance. The sailfish reaches speeds of 68 miles an hour, while the tuna is able to swim at 65 miles an hour. The gentoo penguin's speed of 22 miles an hour seems modest by comparison.

Peregrine falcon
100 miles an hour

Merganser
40 miles an hour

Ostrich
45 miles an hour

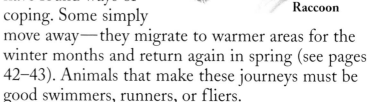

How do animals survive in winter?

In temperate areas of the world, such as Europe and much of North America, animals must adapt to big differences in temperature between winter and summer. Animals which have basked in the summer sun have to face snow, ice, and freezing temperatures in winter.

Mammals and birds need more food in winter to give them the energy to keep their bodies warm. And just when they need more food, it is harder to find it. Snow and ice cover feeding grounds, plants stop growing, water turns to ice, and the soil is rock hard.

Despite all the problems, animals have found ways of coping. Some simply move away—they migrate to warmer areas for the winter months and return again in spring (see pages 42–43). Animals that make these journeys must be good swimmers, runners, or fliers.

Raccoon

Other animals, such as ground squirrels and hedgehogs, hibernate. They sleep through the winter in safe holes, burrows, or caves and avoid the really severe weather. While hibernating, the animal uses as little energy as possible so it can survive on its supplies of body fat. Its body temperature falls and its breathing and heartbeat rates are slower than usual.

Bears, badgers, and raccoons also spend much of the worst of the winter sleeping in warm dens. But they are not true hibernators because their temperature and breathing rate do not change much. In the milder spells of weather, these animals wake up and go out to hunt for any food that is available.

Yet others survive the hard way, out in the winter weather. Their coats get longer and thicker to keep them warm in the cold, and they scavenge for any food they can find. Some sheep and deer scrape beneath the snow with their hooves to reach any available plant life.

Arctic fox
This fox does not hibernate and manages to survive the bitter Arctic winter. Its fur grows thicker in winter to protect it from the cold, and it even has fur on its feet.

WINTER SLEEP

The golden-mantled ground squirrel is a true hibernator. While hibernating, its body temperature falls, and its breathing and heart rate slow down. In this way, the squirrel uses so little energy that it can survive the winter on its stores of body fat. From time to time during hibernation, the squirrel wakes up, and its temperature returns to normal while it is awake.

Golden-mantled ground squirrel

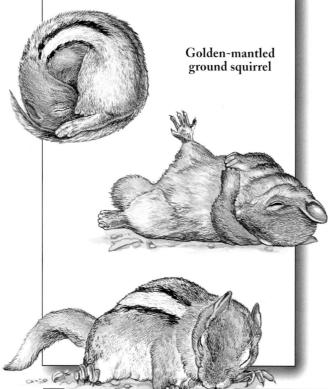

Seasonal lives
Flies die off in fall and winter as the weather gets colder. But they leave their eggs behind them, and enough survive to guarantee plenty of flies the following summer.

Do flies die in winter?

Yes, most do in countries that have severe winter weather. But although the adults die, they leave their eggs to survive through the winter. In this way they make sure that, even though they have died, there will be flies the following year.

Flies breed from late spring through fall. They lay eggs which grow into legless larvae called maggots. These maggots become adult flies. As the winter draws near, nearly all of the flies lay their last eggs and die. Many of the eggs will also perish in the harsh conditions to come. But some that have been laid in well-protected spots usually survive until the spring.

When temperatures start to rise in the spring, the few eggs that have lasted through the winter hatch. Flies reproduce extremely fast, so from these few eggs the population soon builds up again.

There are more than 90,000 different kinds of flies living nearly all over the world. Some, such as midges, mosquitoes, and dance flies, are slender and delicate. Others, such as blow flies and hover flies, have larger, thicker bodies. Most flies have special mouthparts for sucking up flower nectar, plant sap, and other liquid foods. House flies, dung flies, and blow flies also feed on rotting matter such as dung and dead animals. These flies can spread disease as they move from meal to meal.

EGG TO FLY

Flies usually lay their eggs near some suitable food. The eggs hatch into maggots. As the maggot grows, it sheds its skin several times. The last larval skin becomes a hard case called a pupa. Inside, the insect turns into an adult fly.

Eggs

Maggots

Pupae

Fly emerging from pupal case

Adult fly

Which are the most dangerous animals?

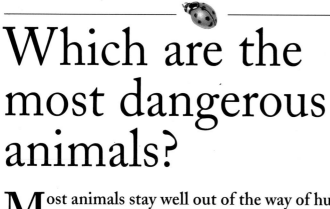

Funnel-web spider

Most animals stay well out of the way of humans if they possibly can and run away rather than attack. However, any large predatory animal is obviously a danger. Sharks are among the most aggressive animals in the sea. Equipped with a huge mouth and rows of jagged teeth, the great white is rightly feared and has killed more people than any other shark species.

Large land animals tend to be most dangerous when protecting their young, and even a normally peaceful creature such as an elephant can become threatening in defense of its family. Among the big cats, tigers are perhaps the most notorious man-eaters. Though few tigers seek out humans to stalk and catch, some will attack on sight—particularly if supplies of their normal prey are low and they are hungry.

Many people are afraid of big snakes, although many species of snake are not poisonous. But not all dangerous creatures are large. Smaller creatures with venomous bites are perhaps even more to be feared than the mighty hunters. The bite of the funnel-web spider can kill a person in two hours, and the tiny arrow-poison frog contains some of the strongest venom known.

Rattlesnake

Arrow-poison frog

SHARKS

There are more than 340 types of sharks, most of which are fierce hunters. A shark's main weapons are its jagged, triangular-shaped teeth which it uses to hold and cut through the flesh of prey.

Hammerhead shark

How long do animals live?

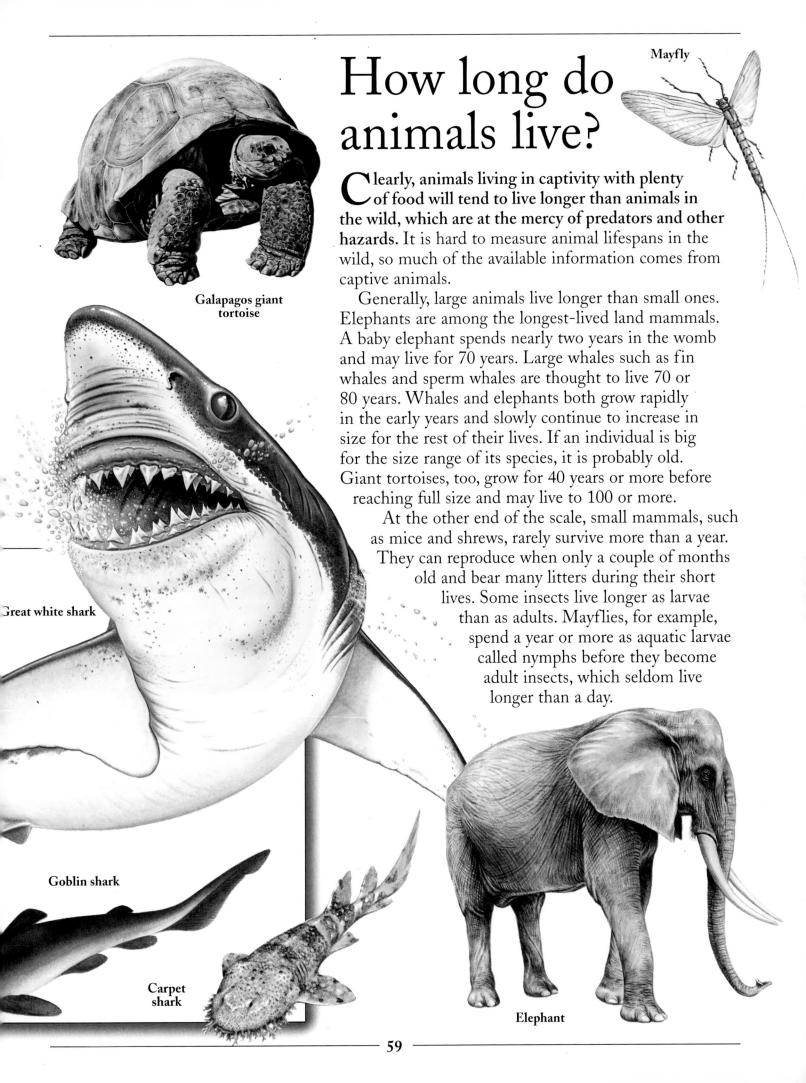

Mayfly

Clearly, animals living in captivity with plenty of food will tend to live longer than animals in the wild, which are at the mercy of predators and other hazards. It is hard to measure animal lifespans in the wild, so much of the available information comes from captive animals.

Generally, large animals live longer than small ones. Elephants are among the longest-lived land mammals. A baby elephant spends nearly two years in the womb and may live for 70 years. Large whales such as fin whales and sperm whales are thought to live 70 or 80 years. Whales and elephants both grow rapidly in the early years and slowly continue to increase in size for the rest of their lives. If an individual is big for the size range of its species, it is probably old. Giant tortoises, too, grow for 40 years or more before reaching full size and may live to 100 or more.

At the other end of the scale, small mammals, such as mice and shrews, rarely survive more than a year. They can reproduce when only a couple of months old and bear many litters during their short lives. Some insects live longer as larvae than as adults. Mayflies, for example, spend a year or more as aquatic larvae called nymphs before they become adult insects, which seldom live longer than a day.

Galapagos giant tortoise

Great white shark

Goblin shark

Carpet shark

Elephant

Why are frogs born as tadpoles?

By spending the first part of its life as a swimming tadpole, the frog gets the best of two worlds. A tadpole is fully equipped for underwater life, with a tail to power its swimming and gills to allow it to breathe in water. It feeds on tiny water plants at first, then on small water creatures.

As it grows, the tadpole gradually turns into a frog and starts a new way of life, breathing air through its lungs and its moist skin. The fully grown frog with its sharp eyes and powerful legs is able to live partly on land catching insects. So for both of the two very different stages of its life, the frog has a body suited to its activities. Because the tadpole and frog lead such different lives, they are not competing with each other for space or for food. Each has its own successful lifestyle.

Do crocodiles lay eggs?

Yes, like most reptiles, crocodiles do lay eggs. The female Nile crocodile digs a nest near water and lays up to 60 eggs which she then covers with soil. For the next three months, she stays near the nest to guard her eggs from hungry predators.

When the young crocodiles are ready to hatch, they call out from inside their shells, alerting their mother who uncovers the nest. If, for any reason, she does not come to open the nest, the young cannot escape and die. The mother then carries the babies in her mouth to a safe stretch of water. She cares for them for several months until they are about 18 inches long and able to catch insects and small fish to eat.

Why does a kangaroo have a pouch?

A female kangaroo has a pouch so that her baby has a safe, warm place in which to develop. This is necessary because the young of marsupial animals such as kangaroos are born after only five or six weeks at the most in the womb, and their bodies are immature.

To prepare for the birth, the mother kangaroo cleans and licks her pouch and belly. This may help her baby find its way to the pouch by smell. A newborn baby kangaroo is only $3/4$ inch long and weighs a fraction of an ounce. With the help of the claws on its front feet, the baby drags itself through the fur of the mother's belly to the safety of the pouch. Once here, it attaches itself to one of the nipples in the pouch. The nipple then swells slightly so that the tiny kangaroo cannot easily lose hold of its food supply.

What is a scavenger?

A scavenger is a creature that feeds on the remains of dead animals. In the tropics large creatures such as hyenas and vultures clear up the remains of animals that have died naturally or have been left by predators. In cooler areas, insects such as flies and beetles are probably the most important scavengers.

Vultures are among the most efficient of the large scavengers. Experiments have shown that most kinds rely totally on sight to find their food. Like most birds, they have almost no sense of smell. Each vulture soars in a large circle, searching the ground below. If a dead or dying animal is spotted, the nearest bird drops to the ground to investigate. Its actions are spotted by other vultures, and they fly over to join in. Within minutes a freshly dead corpse is surrounded by scavenging birds. Vultures usually have bare skin on their heads, not feathers. This is because they often thrust their heads deep into the bodies of dead animals. Head feathers would become clotted with blood, so it is better not to have them.

Why do some animals eat plants and some eat each other?

All animals have to eat some kind of food to provide their bodies with the fuel for life. Plants trap the Sun's energy directly to make the food they need, but animals cannot do that. They can obtain that energy secondhand by eating plants. A plant-eater, or herbivore, is then available to be the food of another creature.

By eating the bodies of herbivores, the flesh-eaters, or carnivores, still gain energy from the sun, the primary source, but they get it thirdhand. This transfer of energy can be thought of as a chain, leading from the plants, which are the first producers, through a series of links, each dependent on the one before it. Plant food is a less concentrated source of nutrients than flesh food. Herbivores, therefore, have to spend much more of their time eating than carnivores do.

For the food chain to operate successfully, there must always be a greater bulk of plants than of animals that feed on them. There must also be far more plant-feeders than flesh-eaters. Because animals eat different kinds of food, a greater number and variety can survive in any area than if all the inhabitants were feeding only on plants.

Why do some animals wake up at night?

Animals that sleep during the day and come out at night do so to avoid the problems of daytime, such as competitors for prey and extreme heat. Owls, for example, come out only at night when their rivals, other flying hunters such as hawks and falcons, are asleep. Owls catch large insects, mice, and other small creatures with the help of their huge eyes and sensitive ears. At night, owls have the skies almost to themselves, except for bats, which hunt for different prey.

Jerboas and many other desert-living rodents spend the day in underground burrows where the air is cool and moist. The fierce daytime heat would harm or even kill them. They only come out at night, when the air above ground is cooler. With their big eyes and sensitive ears, they find food in the dark while avoiding their own enemies.

Why do deer have antlers?

Antlers mark out male deer in the herd. The males use their antlers in fights with each other at the start of the breeding season when they are competing for female deer to mate with.

Antlers grow from the bones of the skull, and the pattern of their growth, shedding, and regrowth is the same every year. Deer usually shed their antlers in late winter or early spring. By the following summer, new antlers have grown and are almost ready for the battles of the breeding season. After the breeding season, muscles squeeze off the blood supply to the antlers. The antlers die and fall off, and the whole cycle starts again.

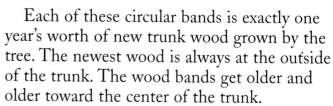

How can you tell a tree's age?

When a tree is cut down, you can see, if the cut has been straight enough, a series of rings in the wood from just inside the bark right to the center of the trunk. Each ring is made up of a double band of light and darker wood stretching right around the tree. The rings are made up of tiny woody tubes, called xylem vessels, that take water and minerals from the roots to the rest of the tree. A new layer of these tubes is always being formed just beneath the bark.

Tree rings
In spring, a tree's growth is fast, and wide tubes are made, the light part of each ring. In winter, when growth slows, thinner tubes form, which are the dark sections of the rings.

Narrow woody tubes

Wide woody tubes

Bark

The oldest trees
Some bristlecone pines began their lives as much as 8,000 years ago.

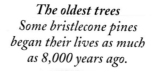

Each of these circular bands is exactly one year's worth of new trunk wood grown by the tree. The newest wood is always at the outside of the trunk. The wood bands get older and older toward the center of the trunk.

One double band means one year of growth. So if you count the number of rings at a point near the base of the trunk, the total will be the tree's age.

A tree does not have to be cut down and therefore destroyed in order to count its rings. You can bore a hole from the outside through to the center and take out a tube of wood with the complete sequence of rings. The rings can then be counted up in just the same way as on the complete trunk.

The oldest trees in the world are the slow-growing bristlecone pines in California. Some of these are over 4,000 years old, making them the oldest living things on Earth. Tree scientists have been able to find out the age of the bristlecones by counting their thousands of tightly packed rings.

Can you tell a tortoise's age from its shell?

Yes, unless it is very old indeed. The shell of a tortoise is made up of a number of differently shaped plates of hard bony material. The number does not increase as the soft body of the animal underneath grows, but the plates do get bigger. If they did not, the tortoise would grow out of its shell!

Just like a tree trunk, the shell plates grow at varying speeds during the year according to

How fast do plants grow?

The speed at which plants grow depends on how much warmth, light, and water they get. With plenty of all three, plants grow very fast. Sunflowers and corn plants can grow from tiny seeds to plants taller than an adult person in under six months. Climbing plants such as the Russian vine are even faster, reaching 20 feet in six months.

All upward growth happens at the very tip of each stalk or twig. Here, at the growing points, called meristems, new cells form.

Sunflower
The fast-growing sunflower can reach a height of 10 feet and has flowers up to 12 inches across.

Wood tortoise
You can tell a tortoise's age by counting up the rings on its shell plates. This method is accurate in a tortoise up to the age of 20 or so, but in older animals the shell is often too worn and battered for the rings to be counted.

weather conditions and food supplies. As a result, yearly growth rings form around each plate. Counting these rings tells you the age of the tortoise. For example, if there are ten rings, the tortoise is ten years old. Some tortoises live to a great age—a hundred years or more—and their shells become worn and scratched. When this happens, it is impossible to see the rings clearly enough to count them up accurately.

Some other types of creatures also reveal their age in growth rings. Bony fish have overlapping scales on their skin which may show annual growth rings. The shells of some sea creatures such as cockles grow throughout the animal's life and show a complex pattern of growth bands.

The green chlorophyll in plant leaves traps energy from the Sun, while carbon dioxide from the air gets into the plant through tiny holes called pores in the leaves. Water from the soil is drawn up into the plant by its roots. Mineral salts such as nitrates, phosphates, and sulfates from the soil are also taken up by the roots and help the plant to build new living matter.

Sunlight

Carbon dioxide

Water

Mineral salts

Do all plants need sunlight?

All green plants need sunlight. Light energy from the Sun is trapped by the green parts of plants and allows them to manufacture the materials they need to survive and grow. This process is called photosynthesis. Plant leaves are light traps: they often grow facing toward the Sun, to catch as much of the life-giving light as they can.

The green in plants is due to a special colored substance called chlorophyll. The chlorophyll is gathered together inside plant cells in tiny green structures called chloroplasts. They absorb sunlight energy and then pass it on through a chemical chain.

This chain uses the trapped energy for two processes. First, it provides the chemical energy to run the plant's living processes. Second, it changes carbon dioxide from the air, and water from the soil, into sugars.

Only a few plants can live without sunlight. Some parasitic plants, for example, are white or cream in color and have little or no chlorophyll. They attach themselves to green plants which can photosynthesize. Using rootlike organs, the parasite sucks the sugars and other foods from the sap of the green plant, stealing its trapped sunlight energy.

Do any plants eat meat?

A few green plants do eat meat as well as making their own food. These plants trap insects with their strangely shaped leaves. Then they slowly digest the insects' bodies and absorb the nutrients. Almost all carnivorous plants live in soils that are poor in

Deadly leaves
Once an insect is imprisoned by the Venus flytrap, its body is dissolved and digested by chemicals inside the leaf.

Pitcher plant
Insects are digested in the pool of liquid at the bottom of the pitcher plant's trap.

nitrogen. They make up their nitrogen shortfall from the creatures that they eat.

Some of the plants have active trapping mechanisms. The large leaves of the Venus flytrap, for example, are fringed with spikes and hinged down the middle. On the surface of each leaf, there are also a number of stiff bristles. If an insect brushes against a bristle, it triggers the trap. The two halves of the leaf snap together and the spikes interlock, imprisoning the insect.

Other plants have more passive techniques. The pitcher plant has flask-shaped structures at the end of its leaves. Insects are attracted to these traps by the sweet-smelling substances just over the rim. When an insect lands on the rim, it falls down the slippery walls of the "pitcher" and is trapped at the bottom. The sundew plant has leaves like flypaper. They are covered with long hairs, each tipped with sticky droplets. When an insect lands on the leaf, it sticks to the hairs. The leaf then folds over around the victim while it is digested.

How do plants grow?

Most plants concentrate their growth at their growing points, the stem tips or root tips. This is quite different from the growth of young animals, who slowly get bigger all over.

The position of a plant's upper growing points, or stem tips (usually found at the centers of buds), decides the overall shape of the plant. If one topmost bud is the main growth point, the plant is a tall, thin one. If growth also occurs from side buds, the plant has a bushier shape.

The underground system also depends on the number of growing root tips. A plant such as a carrot with one main central tip has a strong, deep, single taproot. Where there are several side tips, as in grasses, the root system spreads sideways in the soil.

Each growth tip is made of a cluster of rapidly dividing cells known as meristem cells. These manufacture the new cells that plants need if they are to grow.

Strawberry tree

Flower

Why do plants have flowers?

Plants have flowers to help in the making of new plants. Flowers are needed to make the seeds from which new plants grow. They contain pollen grains and "plant eggs," called ovules. For an ovule to turn into a seed which can be sown in the ground and grow, it must join together with a pollen grain.

Most flowers have both pollen-making and egg-making parts. Ovules are formed in ovaries near the center of the flower. The pollen, which looks like golden dust, forms at the knoblike tips of little threads, called stamens, clustered around the ovaries. Some plants bear two types of flower, one to make pollen and the other to make ovules. However the plant carries its pollen and ovules, the two must get together. Many plants rely on

Strawberry tree fruit containing seeds

The strawberry tree This tree has pink and white bell-shaped flowers which attract pollen-carrying insects. After pollination the seeds develop inside reddish strawberrylike fruits—hence the tree's name.

insects for help. Their flowers contain nectar, a sweet fluid which insects like to eat. Attracted by the flowers' bright colors (see page 68), insects land on them to feed and, while they are there, get covered in sticky pollen. They then carry the pollen from flower to flower, and some grains land on female flowers.

Do all plants have flowers?

Not all plants have flowers. Mosses and ferns, which grow in moist areas almost all over the world, have no flowers. Both reproduce by making spores. Fern fronds have rusty brown spores on their undersides. These fall to the ground and grow into new plants.

Other less familiar flowerless plants include minute plants called algae, which grow as green slime in ditches and lakes, or as a gray-green powdery substance on tree trunks. Seaweed is a bigger form of algae which also never bears flowers. Mushrooms and toadstools are fungi and they, too, do not have flowers. Instead they have tiny spores on their undersides, called gills, which fall off and make new fungi.

Some much bigger plants, such as fir trees and pines, manage to reproduce without true flowers. Instead, these trees carry their pollen and seeds in cones. Large female cones house the ovules, or "plant eggs," while the much smaller, catkinlike male cones contain pollen. Pollen, carried by the wind, joins with the "eggs" which develop inside the familiar woody cones.

Bristlecone pine
The cylindrical cones of the bristlecone pine are up to 2½ inches long. They contain the tree's pollen and seeds.

Bristlecone pine and cone

Bladder wrack
Like many seaweeds, bladder wrack has air bladders like little blisters on its fronds. These help lift the fronds of seaweed apart from each other as they float in water. Bladder wrack lives on rocky shores, and fronds can grow up to 3 feet long.

Which plant has the biggest flower?

The biggest single flower known is the flower of the rafflesia plant, which can be up to 3 feet across. The plant lives as a parasite of a certain type of vine in Southeast Asian rain forests, and the flower is the only visible part. It starts as a bud bulging from the vine and takes many weeks to grow to full size and open out.

The flower has often been said to smell like rotting meat, but in fact it does not smell bad when it first opens. Gradually, though, as the flower begins to decay, its smell becomes more unpleasant. Male and female flowers grow on separate plants, and the rafflesia depends on the many flies which crawl over its huge petals for pollination.

Even bigger than the rafflesia flower is the bloom of the titan arum, which has a spectacular flower spike up to 9 feet tall. But this is a group of many flowers, not a single bloom like the rafflesia.

The world's biggest flower
The rafflesia was first described by Sir Thomas Stamford Raffles, who founded Singapore in 1819. The rafflesia plant was named in his honor.

Foxglove

Why are flowers so pretty?

Pretty, colorful flowers attract insects such as bees and butterflies which, without knowing it, help them in the making of new plants. For a new plant to grow, pollen grains must join with a "plant egg," called an ovule. The pollen fertilizes the ovule so that it can become a growing seed (see page 66). Insects can play a vital role in bringing the pollen and ovule together.

Bright petals and sweet scent advertise the fact that a flower contains nectar, a sweet sugary fluid that insects like to eat. While feeding, the insect gets covered in sticky pollen grains. If the next flower it visits should happen to be of the same type, some pollen grains drop off and are able to join with the ovule. The plant is then fertilized.

The colorful parts of a flower are usually the petals. They must be bright to stand out against the green foliage. The important pollinators, such as bees, flies, and butterflies, usually have color vision—they can tell one color from another. Bees hardly see red at all, but can see ultraviolet light which humans and birds cannot. Bee-pollinated flowers can be almost any color, but red is unusual.

Do only insects pollinate flowers?

Insects, such as bees, flies, and butterflies, are the most common pollinators, but some birds and even bats also pollinate flowers. This happens mostly in the tropics where plants are flowering throughout the year and there is a constant supply of food in the form of nectar.

Flowers pollinated by bats usually open at night and are either large and strong enough to bear the

Nectar-feeding bee
When a bee visits a flower to feed on nectar, pollen grains from the flower cling to the hairs on the insect's body and front legs. On each of the back legs are pollen baskets—special grooves lined with stiff hairs. The bee scrapes the pollen from the rest of its body onto these baskets, which look like yellow blobs when full. Some pollen grains will brush off onto the next flowers that the bee visits, but most of it is carried back to the bee's nest.

weight of these flying mammals, or they are grouped into dense masses. When the flowers open in the evening, bats come to feed on nectar and carry pollen away on their furry heads and necks. Nectar-feeding bats usually have long tongues which they use to reach into deep flowers.

Bird-pollinated flowers are usually red or orange, as birds can see these colors well. Hummingbirds hover in front of flowers to feed (see page 40), but other flower feeders, such as sunbirds and honeyeaters, simply perch nearby. Pollen grains cling to the bird's feathers as it plunges its beak into the flower to feed.

Long-tongued bat

Honeyeater

Flower feeders
The honeyeater lives in tropical forests, where it feeds on ripe bananas as well as flower nectar. The long-tongued bat has tiny bristles on the surface of its tongue which help the pollen to stick fast.

How does a plant know when to flower?

Plants that live for at least a year usually make their flowers at a particular time. Forsythia bushes flower in spring, chrysanthemums in late summer or fall. Flowering starts off the plant's process of reproduction. It must happen at just the right time when conditions are at their best for that particular plant and its seeds have the best chance of success.

The plants' cue seems to be the changing day lengths throughout the year. Green leaves "sense" the changing number of daylight or nighttime hours and from them can tell the time of year—more daylight hours herald spring, fewer hours, fall. The plant can then switch on its process of flower-making at the correct time of year, whether spring, summer, or fall.

New buds

Mark left by last year's leaf

Near a springtime bud of horse chestnut is the horseshoe-shaped mark left by last year's leaf.

Before a leaf falls, a layer of dead corky material forms across the base of each leaf stalk.

Layer of dead corky material

New bud

The stalk and leaf outside the dead layer eventually drop off or are blown away by the autumn winds.

Leaf separating from dead layer

Why do trees lose their leaves in the fall?

Trees that shed all their leaves in the fall do so to protect themselves from the frost and snow of the coming winter. The trees that do this are known as deciduous trees.

Once a tree is leafless, its living interior is sealed off behind the protective corky bark ready for the cold weather. Leaves have tiny holes, called pores, in their surface. If leaves remained on the tree, it would be much easier for ice to form in the interior "plumbing" of the tree via these pores and cause damage. Also, far more snow and ice would settle on leafy branches than on bare ones, and many would break.

Why do leaves turn golden before they fall?

The beautiful color changes of autumn leaves from green through yellow or golden to brown or red are, too, part of the tree's way of protecting itself.

The tree cannot afford to lose the stores of food and minerals held in its leaves, so before the leaves die and fall, the tree must take back anything useful from them. The foods (sugars and proteins) that the leaves have made in summer, by a chemical reaction called photosynthesis, are drawn back into the sap in the tree. (In photosynthesis, plants use the Sun's energy to build foliage from carbon dioxide taken from the air and water drawn up from the roots.)

The leaves change color because the tree also pulls back the green

substance which gives them their color, chlorophyll. This is important in photosynthesis, so it is valuable to the tree.

With the green removed, only yellow and brown colored material, similar to the coloring substance in carrots, remain. These colors were in the summer leaves, but were hidden by the green chlorophyll.

Areas well known for spectacular displays of autumn colors are New England and the eastern edge of mainland China.

Holly

Do all trees lose their leaves?

The leaves on all trees die eventually, but not necessarily all at once. On broad-leaved deciduous trees, such as oak, ash, and elm, all the leaves die in the fall, leaving the tree bare.

A few broad-leaved trees and plants, such as holly, laurel, and ivy, keep their green glossy leaves through the depths of winter. These plants are known as evergreens. The leaves of evergreen plants are especially tough and can stand the winter cold. Their pores close off in winter to protect the inside of the plant, and their glossiness is a waxy layer that prevents them from being damaged by frost.

Even these tough leaves become old and damaged in time and must be shed. They fall in small numbers throughout the year so the tree is never completely bare.

Other trees that remain green all winter are the conifers or cone-bearing trees, such as pines, spruce, and firs. These conifers have tiny leaves like needles which are tough

Monterey pine

enough to stand up to winter cold. Like the holly leaves, the needles fall gradually and are replaced by new ones. Most leaves fall onto the ground above the tree's roots. Here they rot—they are broken down by tiny organisms and fungi in the soil. Some of the nutrients from the dead leaves are taken up again by the tree's roots to help build new leaves. These in turn fall, rot down, and are used again—one of the endless and marvelously economic cycles of plant life.

Spray of pine needles

Towering pine
The Monterey pine grows to a height of about 100 feet. Its long, slender leaves are up to 6 inches long and grow in clusters of three. The tree bears many cones, but they remain tightly closed for many years. They usually open only in the heat of a forest fire.

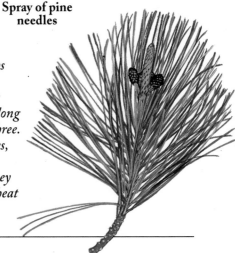

How many types of forest are there?

Most scientists agree that there are three main types of forest. A forest is an area where there are lots of closely growing trees, and the type of forest depends on the climate—the range of temperatures and the amount of rainfall—in the area. The main types are cold coniferous forests, temperate broad-leaved forests, and rain forests.

The cold coniferous forests consist of tall evergreens—trees that have a covering of leaves all year round—such as pines, firs, and spruces. These trees have narrow, needlelike leaves, and cones to protect their seeds. They usually have a tall, pointed, tapering shape—like a Christmas tree. Beneath the main trees are found evergreen dwarf shrubs. Coniferous forests grow where the average yearly temperature is below freezing point and

Pine needles and cone

Oak leaves and acorns

Oak and pine
The leaves shown here are typical examples from deciduous (oak leaves) and coniferous (pine needles) trees.

where it rains little, and only during the short spring and summer.

The temperate broad-leaved forests are found in warmer zones where there is some rain throughout the year. These forests consist mainly of a mix of deciduous trees, whose leaves fall in autumn and winter, such as oaks, beeches, elms, sycamores, and maples. The trees form a leafy canopy, beneath which are shrubs and then a ground-covering layer.

The rain forests, sometimes called jungles, are found in tropical areas near the equator, where there is a lot of rain and temperatures are high (see pages 74–75).

Coniferous and deciduous forest
Two of the three main forest types are shown in these photographs—the snowy landscape and closely packed trees of a cold coniferous forest (left) and the broad, spreading treetops of a temperate deciduous forest in spring.

Are they found in different parts of the world?

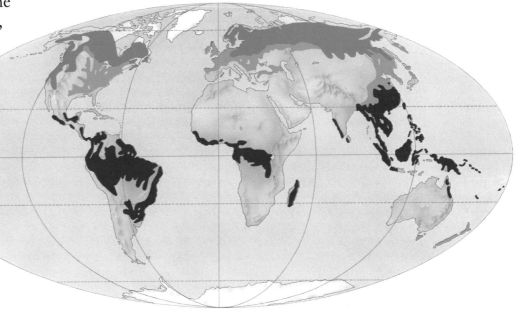

A fire in an Australian eucalyptus forest

Yes, they are. The main types of forest are linked with particular yearly patterns of temperature and rainfall. So the various climate zones around the planet are suitable for the growth of different sorts of forest. The map shows how the main forest types are spread in irregular bands around the world.

Almost all the world's cold coniferous forests are in the northern half of the planet. They form a kind of ring stretching around the world. It starts as an unbroken stripe across North America, from Alaska in the west to Newfoundland in the east. Across the Atlantic, they are found in Scotland and Scandinavia, then in a broad band right across the north of Russia.

By contrast, the tropical rain forests of the world are in three major, separate clumps where the climate is right. These are in South America, centered on the Amazon region; in West and Central Africa; and in Southeast Asia, including Indonesia and New Guinea.

Deciduous forests grow in the temperate regions of the world, where there are cold winters and warm, wet summers. They are found in the middle latitudes of Europe, Russia, eastern North America, and eastern Asia.

Can a forest recover from a forest fire?

Some forests are able to survive and regrow very well after serious fire damage. The gum, or eucalyptus, forests in some parts of Australia are good examples. The foliage, bark, and the leaves that cover the forest floor are so dry that they burn rapidly in a fire, but some well-protected buds can survive. From these, new growth springs after the fire has passed. Some trees and smaller plants in this type of forest even have seeds and fruits which actually need to be damaged by fire before they can start to grow into new plants.

The forests of the world

- Rain forest
- Temperate deciduous forest
- Cold coniferous forest

What is a rain forest?

Rain forest is a type of woodland that grows in hot, wet parts of the world, near the equator. A tropical rain forest, sometimes known as jungle, is the most luxuriant kind of woodland on Earth and contains far more species of plants and animals than any other type of habitat. An area of rain forest measuring only 4 square miles may hold 750 tree species, more than 1,000 types of smaller flowering plants, 400 species of birds, 125 of mammals, at least 150 species of moths and butterflies, as well as thousands of other insect species. Besides the variety, the density of living things in a rain forest is also enormous. Plants and animals breed and grow quickly because of the hot, wet climate.

The rain forest landscape is one of almost unbroken tree cover. More than one and a half million square miles of our planet's surface is covered in this way. Most of the rain forests are in South America, Africa, and Southeast Asia, but there are smaller zones in Central America, the Caribbean, Madagascar, India, and Australia. The tree species, and the animals that inhabit them, differ in each area.

Jungle creatures
Flying frogs live in trees and can glide from branch to branch. The giant anteater and the bushmaster snake stay on the ground, where they find prey. Mandrills sleep in the trees, but find much of their food—leaves, seeds, and insects—on the ground. Toucans feed on fruit in the trees.

Mandrill

Flying frog

Bushmaster snake

Toucan

Giant anteater

What animals live in a rain forest?

Every major group of animals, except those that include only seawater creatures, has species that inhabit jungles. These hot, humid tropical forests contain an astounding number of animal types. Because of the plentiful food supply and wide variety of possible places to live, rain forest mammals, birds, fish, reptiles, amphibians, insects, and other invertebrates abound. Of the 9,000 species of birds that live in the world today, for example, one-fifth of them dwell in the rain forests around the Amazon River in South America.

The animals of the rain forest tend to inhabit a particular forest layer, but are most numerous in the canopy, the main layer of green treetops. They also live in the air above the trees, on the ground, and in the many rivers, lakes, and swamps of these high-rainfall forests. The variety is made greater by the ways in which the lives of animals and plants become linked. Flowers may depend on birds, bats, or insects to carry pollen from one to another (see pages 68–69). Many fruits have attractive colors or smells which draw animals to eat them and so spread their seeds.

What plants do you find in a jungle?

There are all types of plants in a jungle, from mighty trees to microscopic algae that live on the surface of larger ones. There are also many types of fungi on the forest floor.

The variety of plant life in a jungle is staggering. This richness in vegetation is one of the reasons the rain forest is such an important type of ecosystem. It contains a huge treasure-house of plant species that have only just begun to be tapped as sources of human food, medicines, and chemical raw materials.

The tallest jungle trees may grow up to 200 feet high and usually have a single, non-branching trunk, with a broad crown of leaves at the top. Below them, smaller trees form an almost unbroken canopy of green. The lower layers of the forest contain many kinds of

shrubs and an exotic variety of ferns. And the trees themselves act as supports for climbers, vines, and strangler figs. All these plants have roots in the ground which take up water and mineral salts.

In addition, there are a number of other plants which do not have roots in the soil. These include the parasitic plants and the various hanging plants (epiphytes) attached to trees (see page 77).

The colorful heliconia plant grows in the Costa Rican rain forest.

Large-leaved anthurium plants thrive in the high humidity of the rain forest.

Strangler fig
The long roots of a strangler fig wrap around the trunk of the palm tree that supports it.

How do some plants grow without soil?

Hanging plants, or epiphytes, have no roots in the soil; instead they are anchored to the trunks or branches of trees. The whole plant, with its roots, stem, and flowers, is attached to the tree.

Orchids
Epiphytes such as orchids (right) use other plants only as support. They are not parasites.

Epiphytes can probably live in the jungle successfully because the air is so warm and humid all the time. In a typical tropical rain forest, almost all the surfaces of the larger trees are smothered with a profuse growth of these plants.

Jungle bromeliad
A pool of water gathers in the center of a bromeliad flower. Large plants can hold 12 gallons or more.

The most common plants that live this apparently vulnerable type of life are mosses, lichens, bromeliads, and orchids. They gain enough light by growing high above the ground. Their grasping roots mean that they are firmly attached, and they take water directly from falling rain or from water running down a tree trunk. This water can be obtained by dangling roots with absorbent outer coats (a common trick of orchids) or stored as a pool in a cuplike rosette of leaves (a bromeliad method).

Hanging plant
A tillandisa plant clings to its support in a Central American rain forest.

Does anything live at the bottom of the sea?

There are many animals and plants living at the bottom of the sea. What they are and how they live depends on the depth of the sea bottom. In shallow seas close to shorelines, some sunlight filters to the ocean bed. This allows sea-dwelling plants (algae) to live, trapping energy by photosynthesis. These plants are then eaten by ocean animals and form the base of a food chain.

Below about 3,000 feet, the pattern of life changes. Since sunlight cannot penetrate to this depth, there are no living plants on the seabed. Instead of plant life, the foundation of food chains in this part of the ocean is the constant shower of animals and plant remains that fall from the waters above. The bottom-dwelling animals, such as scavenging fish, brittle stars, starfish, sea urchins, lobsters, shrimp, and crabs, depend on this rain of dead and dying food. Other sea animals, such as deep-water sharks, squid, octopus, anglerfish, and viperfish, are deep-sea hunters that capture other animals to eat.

Is it cold at the bottom of the sea?

The deeper parts of the ocean are cold. Just as the sun's light gets absorbed by the water, so, too, does its heat. Below about 3,000 feet, the water temperature near the seabed is only just above freezing, even in the oceans near the equator.

The only exceptions are the deep-sea thermal vent areas. These are found in the darkest ocean depths, near to places where new molten seabed rock is erupting from beneath the Earth's crust. These vents or openings spout hot water that is rich in minerals and in chemicals such as hydrogen sulfide. The water comes from rocks some $1\frac{1}{2}$ miles below the ocean surface.

The area around the vent is much warmer than the rest of the deep sea. In addition, the chemicals in the warm water make it possible for special bacteria to grow which use sulfur, instead of sunlight, to supply them with energy. These bacteria are eaten by other creatures and are the basis of a food chain that includes tube worms, clams, and crabs, which cluster around the vents.

How dark is it at the bottom of the sea?

Below 3,000 feet, the waters of the world's oceans are totally without sunlight. But despite this lack of light, the sea bottom and the deep waters are not completely dark.

A wide variety of deep-sea animals have developed ways of making their own light. There are luminous fish, shrimp, and squid, for example. In some of these creatures, the light is made by clusters of light-making bacteria in the larger animal's body which give off a bluish- or greenish-yellow light. Others make their own "biological" light by a chemical reaction inside the body.

The lights on these animals serve several purposes. For some fish and crustaceans, they are signals that identify the species—the pattern of light allows mates to recognize each other in the dark waters. For other fish, the lights aid their hunting—some deep-sea anglerfish attract their prey by means of a luminous "lure" on the front of the top fin growing from their snout.

Do trees grow on the top of mountains?

The peaks of the highest mountains never have trees growing on them. Part of the way up such mountains is the "tree line," above which no trees can survive. The reason for this has to do with the way trees grow. The growing points of trees are a long way above the ground, at the tips of branches and twigs. In really cold, exposed places, such as high mountain tops, a plant's growth zones would be exposed to the worst of the weather. Tall trees, in particular, are ill-suited to life on the higher mountain slopes because their delicate growing points cannot cope with the cold. In addition, they would not be able to withstand the heavy weight of fallen snow on their slender branches.

On the highest mountains, the upper slopes have no plants living on them at all. This is because they are permanently covered with snow, which lies directly on rocks, with little or no soil covering. Such conditions make ordinary plant life impossible, since all plants need water in liquid form to live and to grow. Although water exists in these places, it is present only as ice and snow. These solid forms of water are no use to plants— they need liquid water which can move through their tissues.

The height of the "tree line" varies in different parts of the world. It is affected by the climate—the amount of rainfall and the temperature levels. In warmer areas, trees can grower higher up a mountain than they can in colder places. In the tropical lands near the equator, for example, lush rain forest covers some low mountain ranges.

Among the plants that are able to live at high altitudes are lichens. Some lichens form thin crusty sheets over the rock surface. They absorb what little water there is from falling rain or melting snow, and minerals from the rock. They can also survive long periods of drying out. In these harsh conditions, they grow very slowly and live for along time. Some mountain lichens grow only $1/32$ inch in a year, but may live for a thousand years.

Mosses, low grasses, and heathery plants also grow at high altitudes. Like the lichens, they can stand the extreme cold and manage to cling to the rocky surfaces.

How high up a mountain can animals live?

Animals can live only as high as there is food for them to eat. In the mountains, high plateaus, and steppe country of Tibet, for instance, there are some large plant-eating animals which can survive in the cold and thin air of the Himalaya Mountains. The two that live highest in this mountain country are the yak and the Chiru antelope, both of which feed on tough, low-growing plants. The yak is found up to 20,000 feet and is well protected from the biting winds and winter snows by its thick fur coat.

The mammals that live highest on some African mountains are the groove-toothed rats and rock hyraxes. On the rock hyrax's feet are flattened nails which look like hooves. Each of these acts like a small, nonslip rubber pad to allow the creature to run easily up the steep rock faces. On the Canadian Rockies, mountain goats wander the craggy slopes in search of the low-growing plants that grow in the brief summer. Marmots, too, eat as much as they can of these summer plants to build up layers of fat to last them through their long winter hibernation.

Large birds of prey soar over some mountains, feeding on both the animal life and the remains of dead creatures. The Andean condor, which flies over the great South American mountain range searching for food, is a typical example.

Where are the world's deserts?

Deserts cover more than a third of the Earth's surface. They are found in North and South America, in Africa, in Central Asia, and in Australia. Deserts are the driest places on our planet and usually have less than 10 inches of rain a year. Some have virtually no rain at all.

What a desert looks like depends on its type of soil, how high up it is, the climate, and the kind of plants that grow there. Where it is depends on the weather pattern. Rain-shadow deserts, for example, are on the sheltered side of mountain ranges, where the mountains act as a barrier against rain. Examples of this type include the Great Basin in North America, the Sahara in Africa, and the Gobi in Asia.

The Great Basin is the biggest desert in the United States. It is high-country desert, most of it about 4,000 feet above sea level, and it stretches between the Rocky Mountains in the east and the Sierras in the west. These two mountain ranges running north to south stop the winds from carrying rain into the desert country. The rain falls on the rising land on each side of the desert, leaving the middle dry.

In the southwestern United States, the Great Basin connects with lower dry regions: the Mojave Desert and Death Valley. Death Valley is so called because of the many early European settlers who perished trying to get through it. Air temperatures as high as 134°F have been recorded there.

Other deserts are dry because they are close to the coast, where cool ocean currents make the winds drop their rain before they reach the land. This is the situation in South America, where a stretch of the western coast fronts a strip of desert called the Atacama, which lies between the ocean currents of the Pacific and the huge mountain wall of the Andes. The Namib and Kalahari deserts in southwest Africa, and the desert in the west of Australia, are made in a similar way—by the cooling effects of an ocean current.

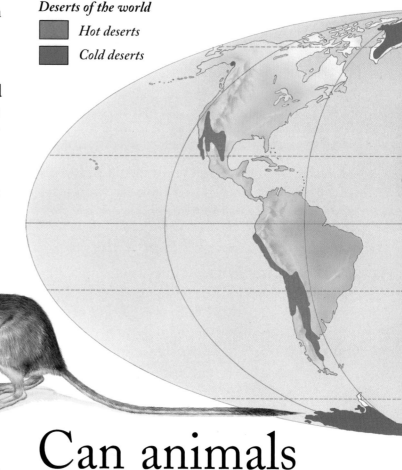

Deserts of the world

Hot deserts

Cold deserts

Burrowing creature
The northern three-toed jerboa lives in the deserts of Central Asia. It spends the daytime sheltering in a burrow which it digs in the soil. At night it comes out to find seeds and insects to eat.

Can animals live in deserts?

Many kinds of animals live in deserts— mammals, birds, reptiles, insects, and other invertebrates. Wherever plants can grow, you will always find animal life as well. The plants of the desert are eaten by desert herbivores (plant-eating animals), and they are, in turn, eaten by several types of desert carnivore (flesh-eating animals). All are adapted to survive in dry conditions.

A camel, one of the largest animals to inhabit the desert, shows many of the special adaptations a creature needs to cope

Gila monster
This lizard lives in the North American desert. It catches birds and small mammals to eat and can store fat in its tail which it helps it to survive when food is scarce.

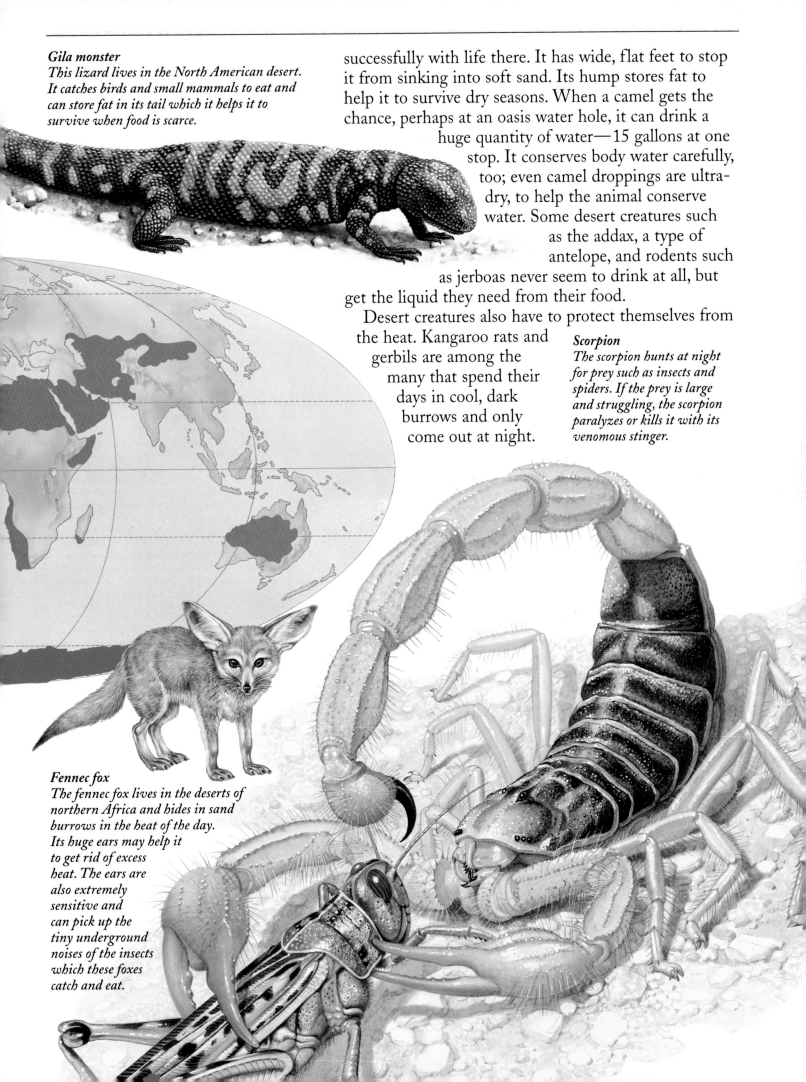

successfully with life there. It has wide, flat feet to stop it from sinking into soft sand. Its hump stores fat to help it to survive dry seasons. When a camel gets the chance, perhaps at an oasis water hole, it can drink a huge quantity of water—15 gallons at one stop. It conserves body water carefully, too; even camel droppings are ultra-dry, to help the animal conserve water. Some desert creatures such as the addax, a type of antelope, and rodents such as jerboas never seem to drink at all, but get the liquid they need from their food.

Desert creatures also have to protect themselves from the heat. Kangaroo rats and gerbils are among the many that spend their days in cool, dark burrows and only come out at night.

Scorpion
The scorpion hunts at night for prey such as insects and spiders. If the prey is large and struggling, the scorpion paralyzes or kills it with its venomous stinger.

Fennec fox
The fennec fox lives in the deserts of northern Africa and hides in sand burrows in the heat of the day. Its huge ears may help it to get rid of excess heat. The ears are also extremely sensitive and can pick up the tiny underground noises of the insects which these foxes catch and eat.

Do penguins only live in cold climates?

Most penguins live and feed in the icy waters of the Antarctic region, but there are a few types that have spread their range farther north.

Penguins are primitive birds which cannot fly. They are adapted for life in the sea and use their flipperlike wings to propel them under water while chasing their fish food. Many types actually breed on the freezing ice shelf of Antarctica itself; others nest on nearby islands. The larger species, such as the emperor penguin, manage to incubate their eggs in the middle of the Antarctic winter when temperatures can fall to −100° F. The male emperor keeps the egg warm on his feet beneath a fold of skin.

The Humboldt penguin is one species that has escaped the harsh climate tolerated by its relatives. It lives and breeds on the western coast of South America and feeds in the rich, fish-filled waters off the coast. The jackass penguin is found on the South African coasts. It comes to land to breed and makes its nest in a burrow or under rocks to avoid the hot African sun.

Emperor penguin

Jackass penguin

What other animals live in cold places?

Animals such as seals, polar bears, and musk oxen, as well as penguins, are all able to cope with the special difficulties of life at low temperatures. Life in a cold climate presents two main problems. First, there is the direct effect of the cold itself. Low temperatures can damage living tissue if a creature's body water starts to freeze. For a warm-blooded animal

ARCTIC ANIMALS

The snowy owl hunts lemmings and any other small animals and birds it can find in its snowy home. Unlike most owls, it will hunt in daylight as well as at night. The walrus is also ideally suited to life in the Arctic and has a thick layer of fatty blubber to keep it warm in the cold sea. It swims well and can push itself along on the ice with its flippers. It feeds on shellfish, which are plentiful in the Arctic.

Snowy owl

Walrus

Tobogganing penguins
King penguins live in Antarctica. Like all penguins, they are superb swimmers but not at their best on land, and they slip and slide as they waddle over the frozen ground. Often the penguins lie down on their bellies and toboggan over the ice—a quicker and easier way to move.

like a bird or a mammal, icy temperatures also mean that more energy is needed to keep up its body temperature. The bodies of animals that live in the coldest parts of the world are well insulated with thick layers of fat or a coat of dense fur or feathers, which stops body heat from escaping too fast.

The second problem is finding food in the snow and ice. These freezing conditions stop most plant growth. They also make it difficult for animals to find prey or to dig food out of soil which has become hardened by ice formation. Penguins, seals, and, for some of the year, polar bears, get much of their food from parts of the sea that stay unfrozen. Caribou dig in the snow using their antlers to find the plants on which they feed. Small animals, such as lemmings, survive by tunneling under the snow to find any seeds and plants that are buried there.

Many of the animals that live in snowy landscapes have white camouflaged bodies to help them merge into the background and prevent them from being easily detected by their prey (see page 50).

ANTARCTIC ANIMALS

It is colder, drier, and windier in Antarctica than anywhere else in the world, making it a very hard place for any animals and plants to live. There are only two types of flowering plant in the whole continent. The only true land animals that can survive on Antarctica are insects, such as lice and fleas, and other tiny creatures such as mites.

Antarctic waters, however, are full of life. There are plenty of fish and other sea creatures, providing food for seabirds and the many kinds of Antarctic seals and whales. Crabeater seals live only in the Antarctic region. They feed mainly on small shrimplike creatures called krill. At more than 100 feet long, the blue whale is the largest animal in the world. Blue whales swim in all the world's oceans, but the biggest population is in Antarctic waters.

Crabeater seal

Blue whale

Do the same animals live in the Arctic and the Antarctic?

Most of the animal species that inhabit the snow-covered lands at the top of the globe are different from those that live at the opposite end of the world. The landscapes are different, too. The Arctic is a huge ice-covered ocean. Land masses such as Greenland stretch above the Arctic Circle, but most of what looks like land is really floating ice. Antarctica, by contrast, is a huge land continent covered in a thick ice sheet.

Polar bears, walruses, and other species of seals particular to the Northern Hemisphere live in the far north of the Arctic. Farther south live ermine, snowy owls, musk oxen, and caribou. There are no large land animals in Antarctica. Penguins and seals spend most of their lives in water and only come to land to breed.

What is tundra?

Tundra is an ecosystem that is found on land where the average temperature is extremely low. Even in the summer, there is still a permanently frozen layer of soil beneath the unfrozen surface of the soil. This perpetual layer of underground icy soil has a special name—permafrost.

Winter temperatures in the tundra zone are well below freezing, and even in the summer months the air temperature is never very high. The rise in temperature, though, is enough to thaw the top layer of the previously frozen ground, allowing some tough, cold-tolerant plants to grow in the thin layer of thawed soil.

Most tundra is in the northern half of the world in northern Alaska, northern Canada, around the coasts of Greenland, and across the top of Scandinavia and the Arctic regions of Russia. North of this Arctic tundra is the ice and snow of the Arctic itself.

Alpine tundra, on the other hand, can form near the top of any mountain where temperatures are cold enough to produce permafrost. This kind of tundra is found in the Alps in Europe, the Rocky Mountains in North America, and even on tropical mountains in the heart of Africa.

Tundra swan

Does anything live there?

Tundra lands are home to a surprising number of animals and birds which are hardy enough to stand the harsh conditions. Animals such as musk oxen and brown bears, for example, are well adapted to living in the tundra and have thick fur coats to keep them warm and dry in winter weather. Smaller animals can, to some extent, escape the worst by living below the winter snow layer. The mice, voles, and lemmings of the tundra often spend the winter months in this way, tunneling under the snow. Hidden from predators, they eat their way through the seeds and other plant matter produced during the previous growing season. Arctic hares also feed on the low-growing plants.

A large proportion of typical tundra species are migrating birds, such as swans, that fly elsewhere to escape the harshest winter months. Snow buntings, too, come to the tundra in summer and nest farther north than any other bird.

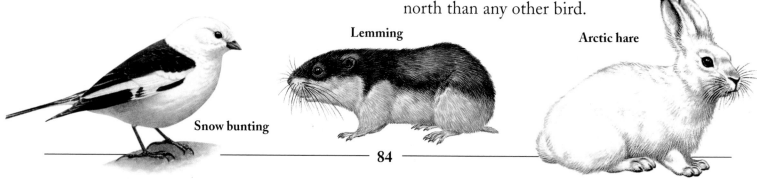

Snow bunting

Lemming

Arctic hare

Tundra landscape
In spring, the tundra snows begin to melt and the top layer of soil thaws. During the brief summer growing season, carpets of small shrubs burst into flower, making a sudden blaze of color in these northern lands.

Are there plants in the Arctic and Antarctic?

A few plants do manage to live in these areas. The freezing temperatures and biting winds in Antarctica mean that only two kinds of flowering plants manage to grow there. These are a type of tough grass and another sort of plant called the pearlwort that forms wind-resistant, low cushions close to the rocky soil. All the other land plants of Antarctica are simple plants such as mosses, liverworts, and lichens which do not have flowers. Lichens are the most successful of these, and there are more than 350 kinds growing in Antarctica.

The plants on the tundra landscapes around the fringes of the Arctic grow quickly in the long days of the brief Arctic summer. Here, there is a wider selection of plant types. Arctic plants include heathers and some dwarf trees only a few inches high, as well as lichens, mosses, and sedges.

Do people live in the Antarctic and the Arctic?

The only people who live in the Antarctic are scientists in research stations. They can only survive there with food supplies and fuel brought in from outside.

The story of people in the Arctic is rather different. The Inuit (Eskimo) tribes use the tundra and the Arctic seas beside it for their survival. They live in the north of Alaska and Canada, as well as in Greenland.

Although most Eskimos now live town lives, in the recent past they were all "hunter-gatherers," able to live off the animal and plant life around them. They killed whales, seals, and walrus at sea, even through the ice, using spears, harpoons, and bows. Musk oxen and caribou provided food on land. Their winter houses, known as igloos, were made of snow.

Fishing through the ice
The Eskimo people are adapted to life in the Arctic.

Alligator

What are wetlands, and how do animals live there?

Wetland creatures
Among the most feared of all wetland animals is the alligator. It lurks among plants in marshes and swamps, waiting for unsuspecting prey such as fish and birds to come near so that it can snap them up in its mighty jaws. Flamingos feed on tiny water animals and plants while herons and jacanas catch small creatures.

Heron

Flamingo

Jacana

Wetlands are the places in the world where open water and land meet. This may be around the edges of lakes, along the sides of rivers, on the banks of estuaries where rivers flow into seas, or along seashores. In these places, waterlogged soils and lots of water make habitats called marshes and swamps. They are both types of wetland.

In different ways, wetlands are like both land and water landscapes. Many sorts of animals and plants live there. The water itself provides a home for large numbers of fish and invertebrate animals; the mud or soft soil is packed with burrowing worms, shellfish, and other animals; and larger mammals, birds, reptiles, and amphibians move among the dense vegetation.

The animals of the wetlands often have special ways of finding food in their marshy surroundings. The flamingo holds its strange-shaped beak upside down in the water and filters out tiny animals and plants. Storks, herons, bitterns, and egrets are all birds that stalk among the reeds and rushes of swamps around the world. They hunt from land or move through the shallow water, holding their powerful, dagger-shaped beaks ready to seize prey such as fish, crabs, small mammals, and frogs.

Jacanas, also known as lily-trotters, catch insects and other small prey on the top of floating water plants. The jacana's toes and claws are so long that its weight is well distributed over a large surface area, and the bird is able to walk safely over floating lily pads without sinking.

How do plants survive in the wetlands?

Each type of wetland has its own particular mix of vegetation, all of it adapted to the marshy habitat in which it lives and grows. Reeds, papyrus, and water lilies grow in an African tropical swamp in rain forest country. An inland bog in a moorland farther away from the equator will have mosses and rushes. Mangrove trees can survive in salty water in swamps at the ocean's edge. They have strong roots which anchor them in the soft mud.

One of the largest wetlands in North America is the Okefenokee Swamp, famous for its peaty brown waters. The water color has developed because dying plant life continually drops into the water and decays, making mats of peat at the bottom of the swamp. Sometimes gases given off as the plants decay force the mats of peat upward. They create floating islands on the surface of the water where plants, and even trees, can take root.

Magpie geese and egrets roam a wetland area in the Northern Territory of Australia.

Okefenokee Swamp
Some of the plants grow on floating mats of peat which have been forced to the surface of the water.

The edge of extinction
In 1982 there were only about 20 California condors left in the wild. In the years since, condors have been reared in captivity with the hope of eventually releasing them to build up the wild population again.

Why do creatures die out?

They die out because they stop being successful enough at breeding. In other words, the adults die at a faster rate than their young are being born. This breeding failure can be due to a change of climate which makes life difficult for the species, the arrival of new, more successful competitor species, the loss of an important food source, or even their whole habitat.

All creatures die out, or become extinct, in the end. Looked at over a long enough time scale, every species develops, survives for a time, and then dies out. The period of survival can be anything from a few hundred thousand years to hundreds of millions of years.

Although extinction is a natural process, the scale and the rate of change has increased enormously this century as landscapes are altered and forests destroyed. It is estimated that as many as 60,000 plant species may be extinct or nearly extinct by the middle of the 21st century if present trends continue.

Have any animals become extinct recently?

In the last few hundred years, animals of all sorts have been dying out increasingly rapidly. The terrible but simple reason for this is that there are too many people on Earth to give all the other animal species enough room to exist.

As human populations have continued to grow (there are now more than five billion of us crammed onto the Earth), we have killed more and more animals for food. Perhaps more importantly, we have taken their habitats from them by turning the natural ecosystems of the world into towns, cities, factories, and farmland. When animals are killed in excessive numbers, or deprived of the habitats that support them, they die out.

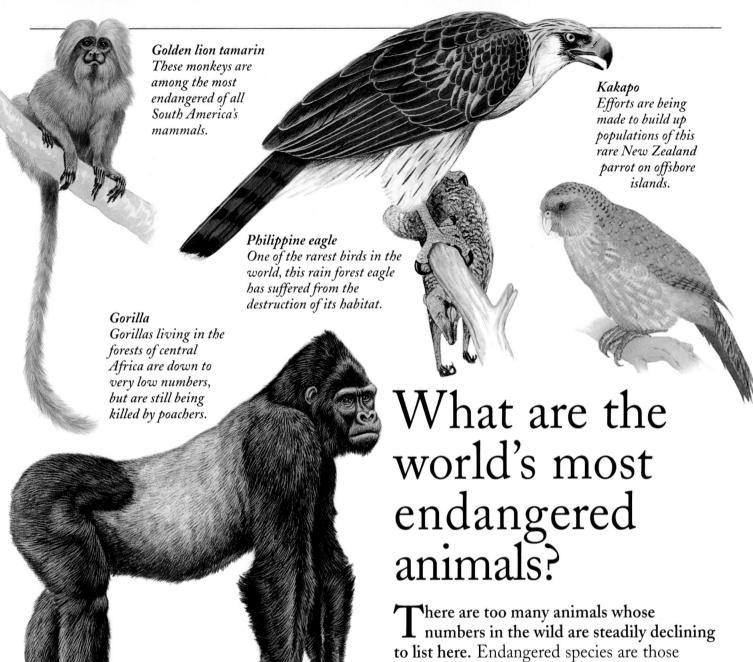

Golden lion tamarin
These monkeys are among the most endangered of all South America's mammals.

Kakapo
Efforts are being made to build up populations of this rare New Zealand parrot on offshore islands.

Philippine eagle
One of the rarest birds in the world, this rain forest eagle has suffered from the destruction of its habitat.

Gorilla
Gorillas living in the forests of central Africa are down to very low numbers, but are still being killed by poachers.

What are the world's most endangered animals?

There are too many animals whose numbers in the wild are steadily declining to list here. Endangered species are those likely to become extinct soon if conservation measures are not taken.

Most of the endangered large animals are in trouble because of human harm. Animals are killed, not only for food but also for other purposes. Crocodiles are destroyed for their skins, elephants for their ivory tusks, and black rhinoceros for their horns, which are used in some eastern medicinal potions. Butterflies have been brought to extinction by over-enthusiastic collectors.

Humans can also cause an animal to become extinct by interfering with its surroundings in some way. Rain forests are being destroyed for agriculture; nesting and feeding sites are being disturbed by people; and poisonous wastes are being dumped in rivers and the sea, killing marine creatures.

Over the last few hundred years, one of the most famous extinctions was that of the dodo in the Indian Ocean. The dodo, a flightless bird living on the island of Mauritius, was relentlessly hunted by humans, and also suffered when pigs were introduced onto the island and destroyed their nests. Steller's sea cow was a large marine mammal which fed on seaweed. It was destroyed by human hunters only 25 years after its discovery in the 18th century. In the 19th century, the passenger pigeon was one of the most common birds in the world. The birds were shot in their millions, and the last passenger pigeon died in the Cincinnati Zoo in 1914.

Our Earth

When a volcano suddenly erupts or an earthquake shakes the ground beneath our feet, it makes us stop and think about the world in which we live. How was it made, how does it work, and how are such dramatic events triggered?

The key to many of the ways in which our planet behaves lies in the fact that the world beneath our feet is not fixed and steady, but is constantly changing. From the gentle ebb and flow of the tides to the gradual weathering of rocks into the soil on which crops are grown, the Earth is in slow but never-ending motion. Below the Earth's solid crust are the invisible movements of molten rocks which cause earthquakes and volcanoes and can also move whole continents.

This chapter takes you on a thrilling trip around the Earth to answer a wide variety of questions about the workings of our unique planet.

The Grand Canyon
The spectacular cliffs of the Grand Canyon contain at least 20 different layers of rock. Each layer reflects millions of years of geological history.

What is under the ground?

The Earth is made up of several different layers. The land we stand on is the Earth's crust—a thin layer like the skin of a peach, made up of the lightest minerals. Beneath this is the mantle, which extends about halfway to the Earth's center. The mantle rock near the Earth's crust is relatively cool and forms a solid region. At greater depths, temperatures are high and the rocks are liquid.

Deeper than about 1,800 miles, a great change takes place, and the mantle gives way to the core. Geologists think that the outer core is probably liquid and the center region solid. The inner core is almost certainly made of iron, mixed with smaller amounts of other elements such as nickel.

Crust
Mantle
Core

The Earth's crust
The Earth has two kinds of crust. The crust which carries the continents is about 22 miles deep on average. The crust under the oceans is only about 3 miles deep.

Crust
Mantle
Outer core
Inner core

Earth's diameter at the equator: 7,926 miles

Earth's crust is about 22 miles deep

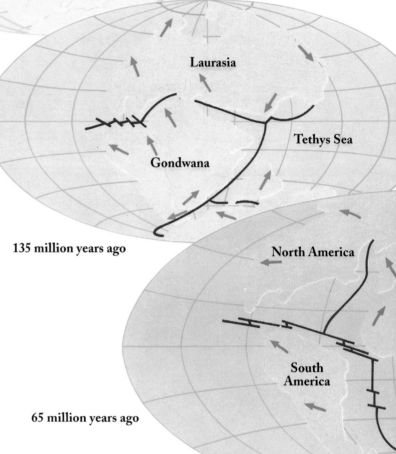

250 million years ago

Pangaea

Tethys Sea

Moving continents
In only about 250 million years, the Earth's landmasses have moved from being one supercontinent, Pangaea, to the arrangement of continents we know today.

Laurasia

Gondwana

Tethys Sea

135 million years ago

North America

South America

65 million years ago

Have the continents always been in the same place?

The shapes of the continents and their position on the Earth's surface have changed dramatically over time and are still changing. North America and Europe, for example, are slowly drifting apart at a rate of about an inch a year—that is, at about the rate that fingernails grow. The North Atlantic gets wider by this amount every year. This rate of movement seems very slow, but it builds up over millions of years.

It is possible for the continents to move because the Earth's surface is not a single unit, but is divided into jagged slabs, or plates. Scientists are still not sure about the plate boundaries in some parts of the world, but they do know that there are seven large plates as well as some smaller ones. All the plates are moving at different speeds toward, alongside, or away from their neighbors. The plates do not move themselves. Instead, they are probably dragged along like giant rafts by currents of heat within the Earth.

Some 250 million years ago, there was one supercontinent called Pangaea (meaning "all land"). About 180 million years ago, Pangaea started to break up as its parts were carried away by plates moving in different directions. At first, two landmasses were formed—Laurasia in the north and Gondwana in the south. By 65 million years ago, Laurasia and Gondwana themselves had started to break up, and the outlines of today's continents came into being.

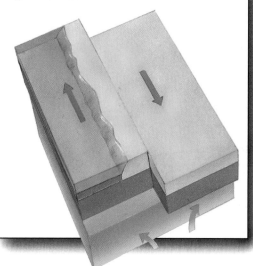

The Earth today
The plates that make up the Earth are shown on these maps. Arrows show the directions in which the plates are thought to be moving.

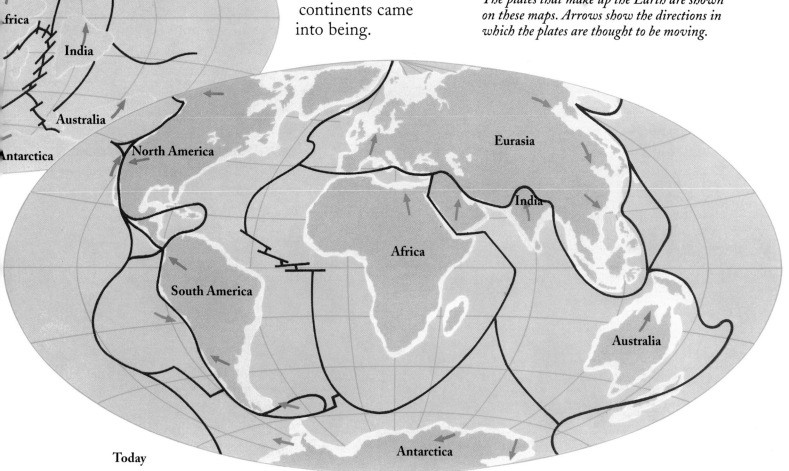

Eurasia

Africa

India

Australia

Antarctica

North America

Eurasia

India

Africa

South America

Australia

Antarctica

Today

93

Are deserts always made of sand?

No, not always. Many deserts appear to be made only of sand, but some have surfaces made of rock, soil, or a stone and sand mixture. This is because it is the amount of rainfall that makes a desert, rather than the type of ground (see pages 80–81).

The Sahara in Africa is the largest hot desert in the world and shows the different types of ground that can exist in a hot desert. It is mainly dry mountain and rocky gravel country, rather than the undulating waves of sand dunes that form most people's idea of a desert.

However, parts of the Sahara really do look like "seas of sand." There are hundreds of thousands of square miles covered with rolling sand dunes.

Types of desert
A hot desert is not always made of sand. Desert landscapes can be made of rock, sometimes worn into strange shapes, rocks and sand, or sand alone. An oasis forms where water in the rocks comes to the surface of the desert .

Sand dunes
Crescent-shaped barchan dunes form when the wind blows from one direction. Seif dunes form when winds blow from two directions. One wind blows across the dune, piling sand up. The other blows along it, moving sand out of the hollows between dunes.

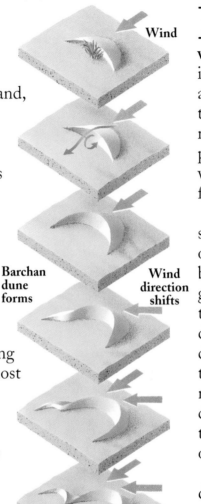

Wind

Barchan dune forms

Wind direction shifts

Seif dune forms

Dune   Oasis   Sandy desert   Stony desert

Does sand in the desert move?

Desert sand is constantly on the move because it is blown by the wind. The wind sculpts the sand into smooth-shaped humps, known as dunes. Though the sand shifts all the time, dunes often form regularly-spaced rows and staggered patterns. These patterns are shaped when the wind blows over the sand from the same direction.

On the windward, gently rising slope of a dune—the one facing the oncoming wind—sand grains roll, bounce, or fly in the wind until they get to its crest. They then fall down the steeper, sheltered side of the dune, where their movement slows down because of the lack of wind there. Since billions of sand grains move like this all the time, the dunes continually lose sand from their windward side and gain it on the sheltered, leeward side.

Another effect of the strong desert winds is that the coarse sand grains gradually wear away even the hardest rocks, forming strange carved shapes.

Are deserts always hot?

Rain

Evaporation

Underground water

Not all deserts are hot. Some, such as the Gobi Desert in Asia, have very cold winters, with snow and freezing winds, and the polar deserts are cold all year round.

Winter and summer are dramatically different in the Gobi. In summer, temperatures soar to 113°F at noon and the ground can be too hot to touch. In winter, strong, chilly winds from the north sweep across the desert, bringing temperatures down to −40°F at times. They also whip up showers of snow which may stay on the ground until spring.

Antarctica and parts of the Arctic are considered to be desert because the rainfall is very low. Antarctica is one of the driest places on Earth. There is plenty of water around, but it is frozen and so cannot be used by animals and plants.

Even hot deserts can be cold at night. There are no clouds in the sky, and the heat escapes so fast after dark that the air may feel very chilly.

DESERT OASES

In some deserts there are pockets of green trees and water, known as oases. Most of the water in an oasis comes from underground and distant water sources. Some oases form where a natural hollow dips into the underground water store. Others are the result of natural cracks or faults in the top layer of rock which allow some of the water in the water-carrying rock below to come to the surface. At these places, water gathers and an oasis eventually forms. Oases become centers for animal and plant life. Date and other palms cluster around the water's edge. Animals such as camels make long journeys to the oasis to find water and food.

The Gobi Desert
Life is hard in winter in the Mongolian part of the Gobi Desert, where people struggle to survive in the freezing temperatures and biting winds.

Wind-eroded mountains

Wind-eroded rocks

Rocky desert

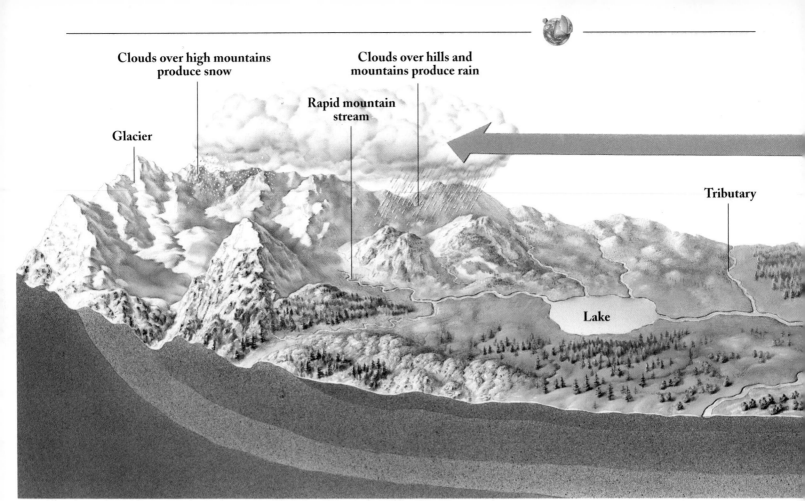

Clouds over high mountains produce snow

Clouds over hills and mountains produce rain

Rapid mountain stream

Glacier

Tributary

Lake

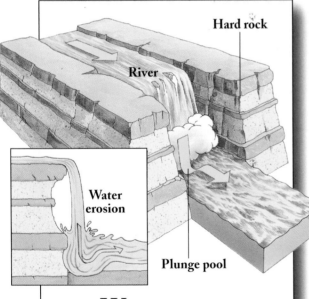

Hard rock

River

Water erosion

Plunge pool

WATERFALLS

A waterfall starts when a river flows over a cliff. Where the water hits the ground, it forms a "plunge pool." Soft rock behind the pool is scoured out. As waterfalls wear away the rock over which they pour, they move back upstream, leaving steep-side gorges behind them. Niagara Falls has moved backward by 1,000 feet in 300 years. All waterfalls are capable of generating lots of energy because the force of water falling over them is extremely powerful.

Where do rivers begin?

Rivers start in high ground where there is lots of rainfall or melting snow or ice. From the hills or mountains of the river's source come the many small, fast-flowing streams that are the beginning of almost all rivers in the world.

Most rivers begin where rainfall is greatest. Water follows a cycle, beginning with the evaporation of water from the surface of the oceans or, to a lesser extent, from lakes, rivers, or vegetation on land, to form clouds. Water evaporates from the seas everywhere, but in greater amounts in the tropics, where temperatures are highest. As the clouds move near mountains, they continue to rise and cool as they travel higher until they release their water content as rain (or as snow if they are cold enough). Rain, snow, and sleet are all forms of "precipitation."

The greatest rainfall occurs on the windward side of mountains, and it is therefore here that most rivers begin. The rain then flows down rivers and streams back to the sea to complete the cycle, as shown above.

The life of a river
Rivers are a vital link in the water cycle between the atmosphere,
from which rain falls, and the oceans, from which water evaporates.
Each river is part of a larger network of streams and tributaries.

Water evaporation
from rivers, land,
and plants

Winding course of
lowland river

River delta

Water
evaporation
from sea

Sea

Which is the world's longest river?

The Nile

LONGEST RIVERS IN EACH CONTINENT

	Miles	Km
Nile (Africa)	4,145	6,685
Amazon (South America)	4,000	6,437
Chiang Jiang (Asia)	3,964	6,379
Mississippi-Missouri (North America)	3,710	5,970
Murray-Darling (Australia)	2,330	3,750
Volga (Europe)	2,290	3,685

The two candidates for the title are the Amazon in South America and the Nile in Africa. Both these rivers measure close to 4,000 miles from their source to where they flow into the sea.

The Nile is generally regarded as the longest river: 145 miles longer than the Amazon when the Amazon is measured along its most direct route. The Amazon, however, carries more water than any other river.

The two rivers are very different. The Amazon's course runs west to east, more or less along the equator—which carries it mostly through tropical rain forests. The Nile flows from south to north, and its course takes it through vast stretches of desert in the Sudan and Egypt where the Nile is the only source of water for farming and irrigation.

Nile

Amazon

Chiang Jiang

Mississippi-Missouri

Murray-Darling

Volga

What is a glacier?

A glacier is a slow-moving river of ice. When rain falls in mountainous country, it runs into streams and river beds and moves swiftly downhill. In cold climates, when snow falls on mountains faster than it melts, it begins to stack up in thick layers. As the weight of new snow presses down, the lower layers turn to ice.

The build-up of pressure eventually makes the frozen mass of ice move down the valley under its own weight. Such glaciers travel downhill far more slowly than any river, normally at speeds of only a few inches a year.

Countryside that once had a lot of glaciers in it has a distinctive appearance. The valleys which glaciers once moved down are usually U-shaped rather than V-shaped and are separated by mountains.

Glacier
A magnificent glacier sweeps across the landscape in the St. Elias Mountains in the Yukon. Most glaciers move only a few inches a day, but some move as fast as 15 feet in a day.

Side glacier

Where do you find glaciers today?

There are more than 4 million cubic miles of glacial ice covering about 11 percent of the world's land surface. Most of this is in the enormous ice sheets covering Antarctica and Greenland, and the rest in thousands of mountain glaciers. There are more than 100,000 mountain glaciers, found on all continents except Australia.

However, during the last cold period of the current Ice Age (see page 25) about 25,000 years ago, glaciers and ice sheets were much larger. They covered not only the continent of Antarctica, but also much of the northern half of North America, most of Great Britain, Scandinavia, northern Europe, and northern Asia. With the coming of the warmer phase of the Ice Age and a milder climate, most of the glaciers over this huge area melted away and disappeared.

A glacier can form wherever it is cold enough to snow regularly. Mount Everest, for instance, is ringed by glaciers because the Himalayan mountain chain is so high and the temperatures so low.

Nose of glacier

End moraine

VALLEYS SHAPED BY GLACIERS

Glaciers change the land they move across. Like a giant bulldozer, the weight of the slowly moving ice, helped by the rocks it carries with it, sweeps away everything in its path. In mountainous country, the powerful scraping by glaciers scours away sharp peaks and steep-sided valleys, leaving rounded summits and U-shaped glacial valleys. Much of the world's most beautiful scenery was sculpted by glaciers during the ice ages. California's Yosemite Valley is a good example of the scenery glaciers can leave behind.

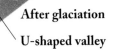

Before glaciation

V-shaped valley

After glaciation

U-shaped valley

Head of glacier

Side moraine

Valley floor

Crevasse

Moving ice
A typical glacier makes its slow descent down a mountainside, sculpting the valley floor as it goes.

How does a glacier move?

A glacier moves under the force of gravity. The movement is like that of a huge conveyor belt, with new ice being added at the top, known as the head of the glacier, and some ice being taken off at the foot as it melts. A glacier moves very slowly. The upper and middle parts move faster than the base and sides, which are slowed down as they press against the floor and sides of the valley.

As it moves, a glacier acts like a huge "file" scraping away and picking up some of the rocks and soil it passes over. Some eroded rocks and soil are carried on the surface of the glacier. When the glacier melts, these may be left in lines called moraines.

A glacier may travel as far as the sea and break into icebergs that drift away. Or the glacier may simply break up when it reaches lower levels, where the climate is warmer. The end of a glacier is called its "nose." Here, the glacier turns into streams and lakes of melted ice.

A glacier's nose
The end, or nose, of the San Rafael Glacier in Chile reaches the sea and the end of its journey.

How did the universe begin?

Most experts think that all matter, space, and time started when something called the "Big Bang" happened about 15 billion years ago. It seems likely that the whole universe started then as a tiny point of infinitely high density and super-high temperature. Only when this point expanded and cooled was it possible for matter, and then stars and planets, to form.

The universe is still expanding as though all its space and matter has been flung apart by some explosion. The echoes of the energy from the Big Bang still exist in space.

How was the Earth made?

The Earth formed at about the same time as the Sun, about five billion years ago. Most scientists think that the whole solar system—that is, our Sun and all its planets, moons, comets, and asteroids—were formed out of a huge cloud of gas and dust. This was made from the dust and debris of an earlier generation of stars.

The forces of gravity squeezed the gases of the cloud together, forming a dense core. The rest of the cloud spun around the core in a flattened disk of gas and dust. Temperatures in the core finally became high enough to cause nuclear fusion. The pressure and heat were so great that the "hearts," or nuclei, of atoms joined together and, as they did, released a great wave of energy. The core had become a star and had begun to push out heat and light—our Sun had been born.

At around the same time, gravity caused the remaining gases and dust particles to form larger and larger rocky masses, which started to orbit the Sun. Some of the larger fragments nearer the Sun became the inner rocky planets—Mercury, Venus, Mars, and Earth. Farther out are the gaseous planets—Jupiter, Saturn, Uranus, and Neptune

What makes the world go round?

The Earth moves in two different ways. It orbits around the Sun—a journey that takes about 365 days—at a speed of 66,500 miles an hour. This orbit is caused by the gravitational attraction between the Earth and the Sun. And it also spins on its own axis, making one complete spin in 24 hours.

The orbit around the Sun is not a perfect circle but a more stretched-out shape called an ellipse. As it orbits the Sun, the Earth spins around an axis which goes through the North and South poles.

How big is the Sun?

The Sun measures about 865,000 miles in diameter and 2.7 million miles around. This gigantic ball of burning gas is by far the biggest thing in our solar system.

The Earth is just a tiny dot in comparison. It would take over a million Earths to make a ball as big as the Sun. Even the total of the nine planets, 43 moons, and all the asteroids (space rocks) in our solar system is only a fraction of the Sun's bulk. The Sun is so large that light, which travels at 186,000 miles a second, takes nearly five seconds to get from one side of the Sun to the other.

Despite its size, the Sun is just a medium-sized star, like other stars in the nighttime sky. It looks different to us on Earth because it is so much nearer to us than any other star.

How hot is the Sun?

The temperature of the surface of the Sun is about 11,000°F. This would melt the hardest of metals.

But the surface is the coolest part of the Sun. The temperature becomes higher and higher toward the center of the Sun until at the center itself it is about 27 million°F. The Sun is the hottest as well as the biggest thing in our solar system. Heat and light from the center are radiated out into space to keep us warm on Earth.

The only time such incredibly high temperatures are even approached on Earth is for a fraction of a second during the explosion of a nuclear bomb.

Why does the Sun rise in the east and set in the west?

The rising and setting of the Sun has nothing to do with the Sun itself. It is caused by the way that the Earth spins.

The Earth orbits the Sun at a distance of about 90 million miles on a roughly circular track. One orbit takes a year. As it orbits, the Earth spins like a top on an imaginary line, or axis, which joins the North and South poles. It spins once every 24 hours, and all the time one half of the Earth is lit by the Sun (daytime), and the other half is in shadow (nighttime). So wherever you live, it is the Earth's spinning that creates the difference between night and day.

If you could look down on Earth from above the North Pole, you would see that the Earth spins counterclockwise—in the opposite direction to the way clock hands move. This counterclockwise movement means that, at any spot on Earth, the Sun first appears over the eastern horizon at "dawn" and disappears over the western horizon at "dusk."

What are sunspots?

Sunspots are small dark patches that show up against the Sun's glowing surface. The Sun's outer layer is a brilliantly white surface with a temperature of about 11,000°F. The darkness of the sunspots is a result of their coolness. They are at a temperature of only 7,200°F!

Scientists think that sunspots are caused by the magnetic activity within the Sun. There is a regular cycle to the appearance of sunspots, with peaks every 11 years.

Which are the largest and the smallest planets?

By far the largest planet in our solar system is Jupiter. It is made up of the gases hydrogen and helium, and so it does not have a solid surface like a rocky planet. It has a diameter of about 86,000 miles and is 300 times larger than Earth.

Mercury, the closest planet to the Sun, is only 3,000 miles in diameter. Pluto, the outermost planet of the solar system, is even smaller, but it is so far away that scientists have only recently been able to measure it. With a diameter of only 1,454 miles, Pluto is smaller than our Moon.

What causes flooding?

Flooding is caused by an excess of water on land, which is sometimes unexpected and sudden. Floods may happen when a river overflows its banks, often due to very high rainfall. Trees slow down the rush of rainwater into the rivers, so river flooding is made worse if nearby forests have been cut down. Seawater flooding can be caused by high tides, or by tidal waves created by earthquakes, the eruption of volcanoes under the sea, or severe storms.

Floods can sometimes be helpful rather than harmful. Before Egypt's Aswan Dam was built, for instance, the yearly floods of the River Nile were an important part of the Egyptian agricultural system. Those flood waters, caused by increased rainfall at the source of the River Nile in Ethiopia, brought water and rich, fertile mud to the Nile's flood plain and its delta. People relied on both the water and the river-borne mud for growing crops by the river. The effect of building the dam has been to even out the river's flow through the year.

Most floods, though, have disastrous consequences. They may drown people and animals, make people homeless by destroying their houses, and ruin their farmland and crops.

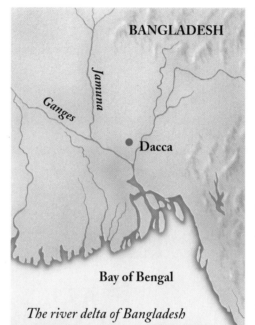

The river delta of Bangladesh

Flooded streets
Exceptionally heavy rains in Illinois caused widespread flooding and forced people to leave their homes.

Which countries have the worst floods?

The worst floods are those that cause serious damage and kill people. They are most common in parts of the world, such as Asia, where the weather conditions make flooding likely. These countries also have very large populations.

The rivers, river deltas, and coastlines of Southeast Asia are particularly prone to flooding. Huge storms at sea can create high tides and swollen rivers, and this happens especially around the large river delta system that makes up much of the coastline of the country of Bangladesh.

Bangladesh is one of the poorest and most overcrowded countries in that part of the world. Many people live at great risk around the river delta, on shallow mud islands only a few feet above sea level. In a recent flood disaster, the death toll reached hundreds of thousands. Terrible river-flood disasters have also occurred in China.

When do monsoons happen?

Monsoon is the name given to the heavy rains that fall at particular times of the year in Asia and parts of Africa. It comes from the Arabic word meaning "fixed season."

This seasonal pattern of rainfall is linked to the tilt of the axis around which the Earth spins, and to the heating up of the tropical areas by the Sun. The region near the equator receives most heat from the Sun, which causes an area of low pressure (hot rising air) at the equator all around the globe. The summer monsoon is caused by the area of low pressure that develops over southern Asia as the land warms up.

In the northern hemisphere's summer, the Earth's tilt brings greater heat and more stormy weather systems north of the Equator. These changes carry the monsoon to India and Southeast Asia, from May to September.

City rains
The seasonal monsoon rains can cause severe flooding in the streets of Calcutta in India.

In winter the rain-carrying winds reverse and bring rains to areas with northeastern coastlines, such as Indonesia and Australia. The monsoon rainfall is both a blessing and a problem for the countries it affects. It is an important source of water for irrigation in hot, tropical conditions, but can also cause flooding and serious damage to homes, roads, and farmland.

PEAK SEASON RAINFALL

JUNE–AUGUST

Cherrapunji, Bangladesh	300 inches
Bombay, India	88 inches
Douala, West Africa	80 inches
Hong Kong	46 inches

DECEMBER–FEBRUARY

Surigao, Philippines	62 inches
Darwin, Australia	40 inches

Summer tropical rain area

Winter tropical rain area

The monsoon
This map shows changes in the position of the monsoon through the year.

Equator

Cherrapunji Hong Kong
Bombay
Douala
Surigao
Darwin

How are mountains made?

A mountain is a steep, rocky mass that rises more than about 2,000 feet above the surrounding surface of the Earth; anything lower is a hill.

Most mountain ranges are made when rocks are pushed upward by movements of the Earth's crust. This uplift happens when continents bump into one another, and one part of Earth's crust slides under another. The highest mountain range of all, the Himalayas in Asia, was formed when the continental plate carrying India moved north and collided with the Asian plate (see pages 92–93), forcing land upward.

Individual mountains can be formed by volcanic activity which continues over tens of thousands of years in one spot. The erupted lava builds up over the erupting center to form a cone of rock. Etna, a mountain in Sicily which is about 9,000 feet high, has been built by volcanic activity over the last two million years.

Mountain peaks
Mount Moran, in Grand Tetons National Park in Wyoming, mirrored in the Snake River.

Caves
This network of underground caves and passages has been carved by the action of water.

Rockfall

Stream

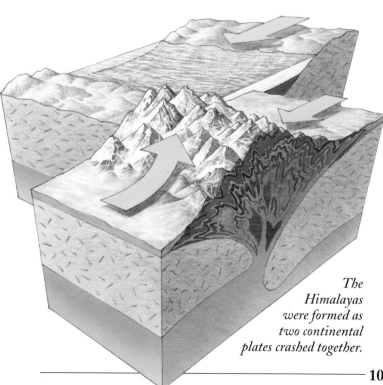

The Himalayas were formed as two continental plates crashed together.

How are caves made?

A cave is formed when a large, hollow space is eroded from rock. Sea caves are usually formed by the pounding action of the waves. Inland caves are made by the chemical action of rainwater.

Sea caves are common where strong waves crash against a rocky shoreline, particularly if the cliffs are made of rocks of differing hardness. Waves are very powerful, and their pressure can wear rock away. This happens most often when the waves carry sand and pebbles. Under such attack, softer parts of the cliff face wear away faster than harder parts, and caves may be created.

Inland caves can form in several kinds of rock, but are most common in limestone. The gas carbon dioxide, which is found in our atmosphere, dissolves in rainwater, which becomes very dilute acid. When the rainwater falls on limestone, it can dissolve the rock. As rainwater trickles under the ground along cracks in rock, it can, over thousands of years, cut huge caves.

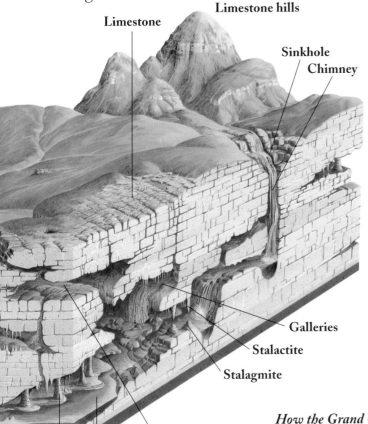

Limestone hills

Limestone

Sinkhole

Chimney

Galleries

Stalactite

Stalagmite

Dry cave

Underground lake

Column

How does a canyon form?

A canyon is a particular type of steep-sided valley that only occurs in special conditions. Canyons form where a river fed from abundant rain or snow flows steadily through dry country. The land is continually being uplifted to keep pace with the down-cutting of the rivers. This means the river can cut a slot-shaped canyon valley through the bedrock. Because there is hardly any run-off of water into the canyon, its sides do not get worn away by water, so they remain almost vertical. Such rocky, desert conditions occur in certain parts of the United States, such as Arizona and Nevada.

Located in northwestern Arizona, the Grand Canyon is the largest and most famous canyon in the world. It was cut through desert rock by the torrential force of the Colorado River. The Grand Canyon started to be cut about 26 million years ago, when the Colorado Plateau, of which the desert is part, was slowly pushed upward by movements of the Earth's crust. As the land slowly rose up, the Colorado River continued to cut its way down. Now the deepest part of this great canyon is about 7,000 feet below the desert plateau.

How the Grand Canyon was formed
Millions of years ago, shallow seas deposited layers of sandstone, shale, and limestone on top of ancient rock. Movements of the Earth's crust created huge block mountains which were worn flat by weathering. The Colorado River then began to cut its way into the layers of rock, creating the canyon.

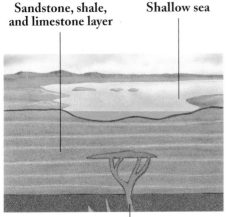

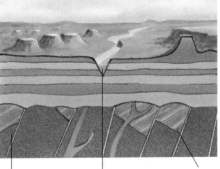

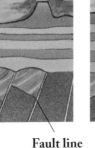

Sandstone, shale, and limestone layer

Shallow sea

Canyon carved by river

Granite intrusion

Block mountains

River begins to form canyon

Fault line

Why is the Earth round?

The Earth is almost round because, when it was very young, the forces of gravity pulled it into a ball-like shape. When you look at the rigid rocks of mountains, it is difficult to imagine that anything would be powerful enough to mold the Earth into a ball. But most of the Earth's huge mass was in fact made of molten, or melted, rock, which is a liquid. It is easier to see how that liquid might have formed into a rounded shape, like a drop of oil in water.

Our planet was formed over four billion years ago. Heat generated in the rocks caused most of its rocks to melt, and gravity pulled the Earth into an almost perfect sphere, more than 3,750 miles in diameter.

The molten rocks eventually sorted themselves into layers, with the densest, or heaviest, settling in the middle of the Earth.

The Earth from space
A computer-colored image of the Earth taken by the Meteosat satellite.

How are islands formed?

An island is a piece of land completely surrounded by water. It may be no bigger than a large sand bank, or hundreds of miles across, like Britain, or even an island continent such as Australia.

Islands are made in four main ways. If the sea rises, it can turn what used to be part of a landmass into an island by cutting it off. Or, the land may rise up above sea level, usually as a result of volcanic activity, to make a new island. Part of a continent may also move away from the rest of it and form a huge island. Greenland and Madagascar are examples of islands formed in this way.

A final way some small islands are formed is from coral reefs (see pages 46–47).

The Isles of Scilly, off the southwest coast of England, were formed when the sea level rose. These granite islands were joined to Cornwall, the westernmost tip of England, during the last Ice Age, when sea levels were lower than they are today because so much water was frozen and locked up as ice. When

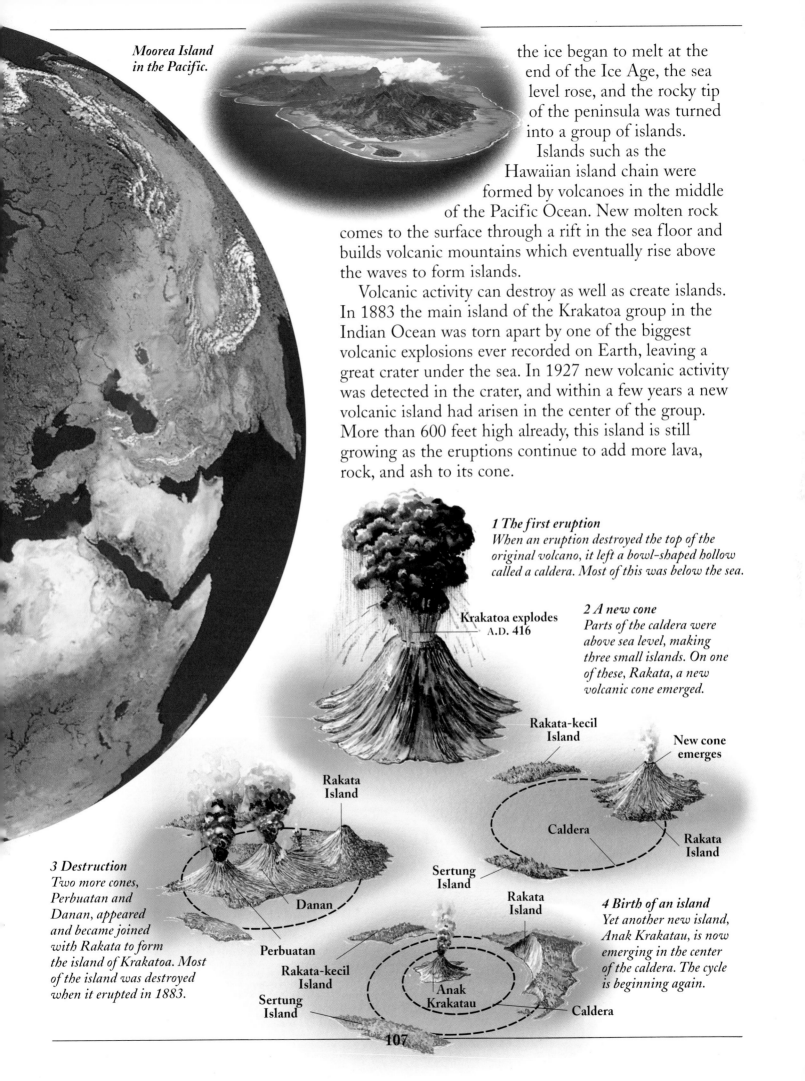

*Moorea Island
in the Pacific.*

the ice began to melt at the end of the Ice Age, the sea level rose, and the rocky tip of the peninsula was turned into a group of islands.

Islands such as the Hawaiian island chain were formed by volcanoes in the middle of the Pacific Ocean. New molten rock comes to the surface through a rift in the sea floor and builds volcanic mountains which eventually rise above the waves to form islands.

Volcanic activity can destroy as well as create islands. In 1883 the main island of the Krakatoa group in the Indian Ocean was torn apart by one of the biggest volcanic explosions ever recorded on Earth, leaving a great crater under the sea. In 1927 new volcanic activity was detected in the crater, and within a few years a new volcanic island had arisen in the center of the group. More than 600 feet high already, this island is still growing as the eruptions continue to add more lava, rock, and ash to its cone.

Krakatoa explodes
A.D. 416

1 The first eruption
When an eruption destroyed the top of the original volcano, it left a bowl-shaped hollow called a caldera. Most of this was below the sea.

2 A new cone
Parts of the caldera were above sea level, making three small islands. On one of these, Rakata, a new volcanic cone emerged.

**Rakata-kecil
Island**

**New cone
emerges**

Caldera

**Rakata
Island**

**Sertung
Island**

**Rakata
Island**

3 Destruction
Two more cones, Perbuatan and Danan, appeared and became joined with Rakata to form the island of Krakatoa. Most of the island was destroyed when it erupted in 1883.

Danan

Perbuatan

**Rakata-kecil
Island**

**Sertung
Island**

**Rakata
Island**

**Anak
Krakatau**

Caldera

4 Birth of an island
Yet another new island, Anak Krakatau, is now emerging in the center of the caldera. The cycle is beginning again.

How much of the Earth is covered in water?

More than two-thirds of the surface of our planet is covered by water, and nearly all of it is in oceans. The total amount of seawater on the Earth's surface is estimated to be 325 million cubic miles—that is something like 100 billion gallons of seawater for every person on Earth. The largest body of seawater is the Pacific Ocean, which has an area of about 64 million square miles. The other oceans in order of size are the Atlantic, Indian, Antarctic, and Arctic.

But, because all of this huge amount of water has salt dissolved in it, we cannot drink seawater, or use it to irrigate our fields.

Watery world
The oceans cover about two-thirds of the Earth's surface.

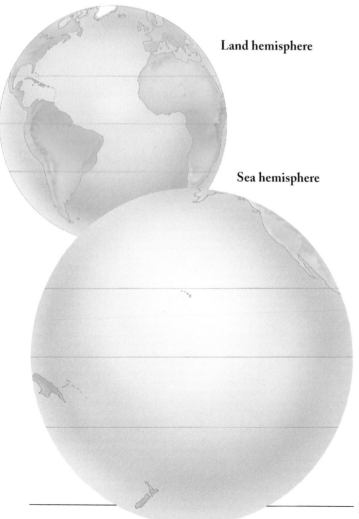

Land hemisphere

Sea hemisphere

What makes the sea salty?

Salt! The salts in the sea have been washed out of the Earth's rocks and soils by rainwater and carried in streams and rivers down to the sea. The proportion of salt dissolved in seawater varies slightly from ocean to ocean but is between 33 and 38 parts per thousand by weight. A large amount (about 85 percent) of these salts consists of common salt (sodium chloride), the type we put on our food. In the remaining 15 percent are other salts and many elements, even gold, but in incredibly small amounts.

To demonstrate the saltiness of the sea, if you boiled a pint of seawater in a saucepan until all the water evaporated, it would leave two-thirds of an ounce of solid salt, mostly common salt, behind.

Sodium chloride is an essential part of our diet. Other mineral salts in the sea, such as those of phosphorus and nitrogen, are crucial for the survival of tiny plants, such as the plankton on which many small fish feed.

Salt works
A pile of gleaming white salt at one of the world's largest salt works in Baja California, where huge shallow pools of seawater are left to dry in the hot sun.

How do we obtain salt?

The salt we use on our food comes from many different places around the world. The main source is the sea itself. In hot countries, shallow pools of seawater soon dry up in the sun to leave a thin layer of solid sea salt on the ground. Another important source is from underground layers of salt rock, which can be mined.

Saltpans are flat, white expanses of dried salt found in hot, dry landscapes such as deserts. The salts involved are those which are found in soils and rocks—mainly a mixture of chlorides and sulfates. The "pan" forms because of the heavy rains that occasionally fall in desert areas. These wash salts out of the soil and rocks of higher ground and carry them in temporary, steep-sided streams (called "wadis" in Arabic) to the lowland areas. If there are suitable hollows in the ground, shallow lakes form. These quickly lose water through the fierce evaporation that takes place under the hot desert sun. Soon the water turns to a salty slush, and then the surface dries completely to form a saltpan.

What is a lake?

A lake is a large, water-filled dip in the Earth's surface. Most lakes contain fresh water, but there are some saltwater lakes such as the Caspian Sea, the largest lake in the world (see page 118).

Lakes are formed in many different ways. A rift valley lake forms where movements of the Earth's crust have left a deep chasm which is then filled with water. A crater lake occupies the hollow left by an extinct volcano. Oxbow lakes can be made when a twisting bend in a lowland river gets cut off from the rest. The power of desert winds can scour out a hollow in the ground. If this reaches underground water, a deflation lake forms. A tarn lake occupies a deep rounded basin made by the scooping action of a glacier.

Types of lakes
Some of the natural ways lakes are formed are shown below. Artificial lakes are made by damming the flow of a river.

| Rift valley lake | Crater lake | Oxbow lake | Deflation lake | Tarn | Artificial lake |

Where does the sea go when the tide is out?

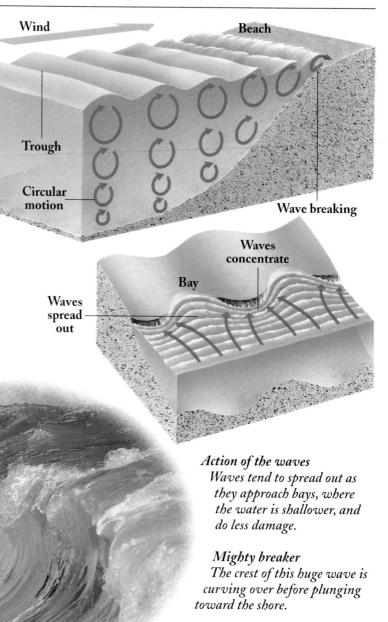

Wind

Crest

Trough

Circular motion

Beach

Wave breaking

Waves concentrate

Bay

Waves spread out

When the tide goes out, the water does not disappear—it just moves from one place to another on the Earth's surface. This happens because the ocean's waters are pulled toward the Moon by gravity.

Gravity is the force that makes objects fall to the ground when you drop them—the Earth's gravity pulls the objects toward it. In the same way, the Moon's gravity pulls the oceans toward the Moon. The part of the ocean directly under the Moon is heaped up into a bulge of water. There is a similar bulge on the opposite side of the Earth. On the coasts of the "bulge" areas, there are high tides. In between, where water levels are lower, there are low tides.

In each period of 24 hours, everywhere on Earth passes though two water bulges and two lower levels—that is, two high tides and two low tides.

Action of the waves
Waves tend to spread out as they approach bays, where the water is shallower, and do less damage.

Mighty breaker
The crest of this huge wave is curving over before plunging toward the shore.

How are waves formed?

Waves are the moving ridges of water created when a wind blows in the same direction for some time over the sea.

The surface of the water puckers into ridges (crests) and hollows (troughs).

The lines of crests are roughly at right angles to the direction the wind is blowing. They move in the same direction as the wind. But, despite all the apparent movement, the water in a wave simply moves up and down in a circular path, without advancing very far.

Waves only "break" when sudden strong gusts of wind push the crest tops over, creating foam-tipped "white caps," or when the waves reach the shore. When a wave approaches land, the bottom of the wave is slowed down by its contact with the seabed. As the top moves ahead, in front of the wave's sluggish base, it reaches a point close to the shore where it falls forward and breaks. This is how the surf forms.

Do waves wear away the coastline?

The action of waves and currents in the sea constantly wears away parts of coastlines. The waves of the sea are a powerful destructive force, especially during storms. Their ability to wear away the shoreline is easily seen along cliff coasts where the rocks are relatively soft. The waves constantly battering at the cliff base cut the rock away, leaving caves, arches, and stacks. This is called coastal erosion.

Rocks that fall into the sea when cliffs are worn away are pounded by the waves and broken down into smaller and smaller stones, pebbles, and gravel. These stones become smooth and rounded by being rubbed and jostled together. Some types of rocks, such as sandstone, break down into minute fragments called sand grains.

The material worn away from the cliffs, together with the large amounts of sediment swept onshore from the seabed and dumped by rivers, moves along the shore to be deposited as beaches, sand spits, and sandbars. This is called coastal deposition.

Where the waves are weaker and the rocks harder, there is less coastal erosion and the coastline is changed more slowly. Spits and sandbars may develop across river mouths, making lagoons. These eventually become filled in with mud and silt from streams flowing into them.

Where a wave breaks at an angle against a steep beach, its upward movement, or swash, drives sand and pebbles diagonally up the beach (see bottom diagram). The return flow, or backwash, drags sediment down the beach at right angles to the shoreline. As more waves come in and break, they move the sediment along in a zigzag pattern.

The changing coastline
The wave action of the sea (along with sand and pebbles carried by the waves) wears away coastlines, creating a variety of coastal forms.

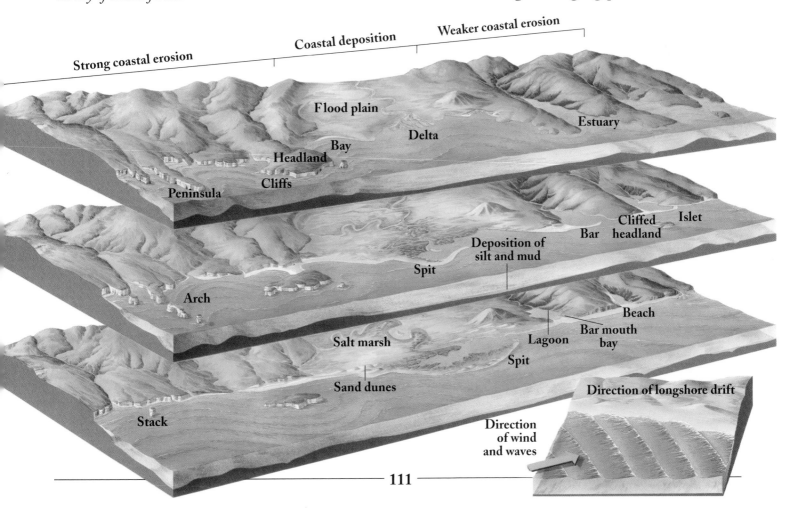

Strong coastal erosion   Coastal deposition   Weaker coastal erosion

Flood plain

Bay

Headland

Cliffs

Peninsula

Delta

Estuary

Arch

Spit

Deposition of silt and mud

Bar

Cliffed headland

Islet

Salt marsh

Lagoon

Spit

Beach

Bar mouth bay

Sand dunes

Stack

Direction of wind and waves

Direction of longshore drift

What is an earthquake?

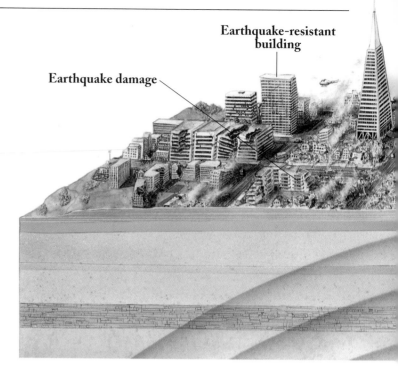

An earthquake is a violent release of pent-up energy in the Earth's crust. In a few seconds, huge forces that have built up slowly over many years in the rock of the plates covering the Earth are released as these rocks adjust to the slow movement of the plates.

The power of these events is enormous, and there is enough energy to send out vibrations or waves through the whole Earth. Near the earthquake we feel these tremors, and they can cause shaking and much damage to property. Farther away from the earthquake, the intensity of the shaking is less, although it can still be sensed by special instruments and recording equipment.

The place where the release of energy begins is called the earthquake's focus. It is always underground in the plates covering the Earth and may be up to 450 miles deep. The point on the Earth's surface directly above the focus is known as the epicenter. Major earthquakes can also open up large cracks in the ground as the rocks become compacted by the shaking. During serious earthquakes, huge pieces of the land slide past one another, or move up or down. These effects are sometimes permanent.

What causes an earthquake?

Earthquakes strike when the huge pressures that build up deep inside the Earth become greater than the strength of the rocks. As a result, the rock becomes strained and breaks.

Where the rocks are weak, continual tiny breaks release the stress. We may see that the rocks have slipped, but there has been no earthquake. Where the rocks are stronger, the pressure builds up until eventually the rocks break in an earthquake. The longer the time interval over which pressure accumulates, and the greater the energy stored in the rocks, the larger will be the break and the greater and more powerful the earthquake.

There are about 500,000 earthquakes every year, but some are so small that they are barely noticed. Others cause great shock waves that echo round and round inside the Earth for a day or more.

Earthquake damage
Deaths during earthquakes are usually caused by buildings, which have been badly constructed or built on unsuitable foundations, falling on people. The most dangerous earthquakes are those that happen in densely populated areas such as towns and cities.

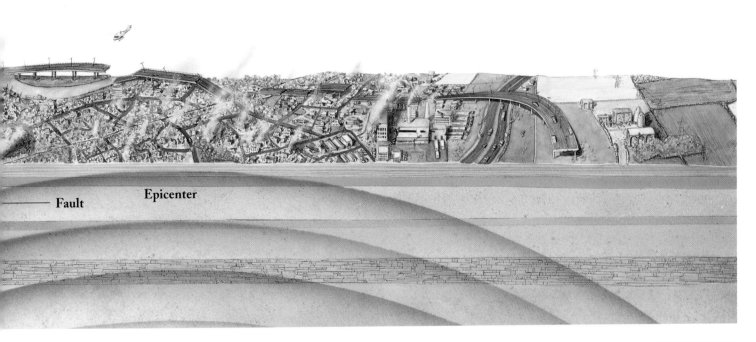

Fault

Epicenter

Seismic waves
Two types of vibrations, or seismic waves, radiate from an earthquake's focus. Body waves travel inside the Earth, spreading out in all directions. Surface waves travel on the Earth's surface away from the epicenter and usually cause more damage.

Focus of an earthquake
Vibrations from an earthquake start at the focus and radiate out in circles, like ripples in a pond. The vibrations may be felt as far as 250 miles away. The nearer the focus is to the surface, the greater the damage tends to be.

Epicenter

Edge of crustal plate

Fault line

Focus

How are earthquakes measured?

An earthquake can be measured in two ways, which tell us different things about the force of these terrible natural events. First, the physical effects of the earthquake can be measured by looking at the degree of destruction and change. This scale, called the Mercalli Scale, has 12 points, ranging from Level I (an earthquake only slightly felt) to Level XII, in which utter destruction is caused.

The Richter Scale uses vibration meters called seismographs to measure the total amount of energy released by an earthquake. These meters can record shock waves from an earthquake even when they are hundreds of thousands of miles from its epicenter. The biggest earthquakes ever recorded measured in the range of 8–8.5 on the Richter Scale. The terrible San Francisco earthquake of 1906 reached 8.3.

MAJOR EARTHQUAKES OF THE 20TH CENTURY

1908	Messina, Italy	killed 110,000
1920	Gansu, China	killed 200,000
1927	Qinghai, China	killed 200,000
1935	Quetta, Pakistan	killed 30,000
1939	Erzincan, Turkey	killed 32,000
1970	Northern Peru	killed 67,000
1976	Guatemala	killed 23,000
1976	Tangshan, China	killed 655,000
1988	Armenia	killed 25,000
1990	Western Iran	killed 50,000
1995	Kobe, Japan	killed 5,200

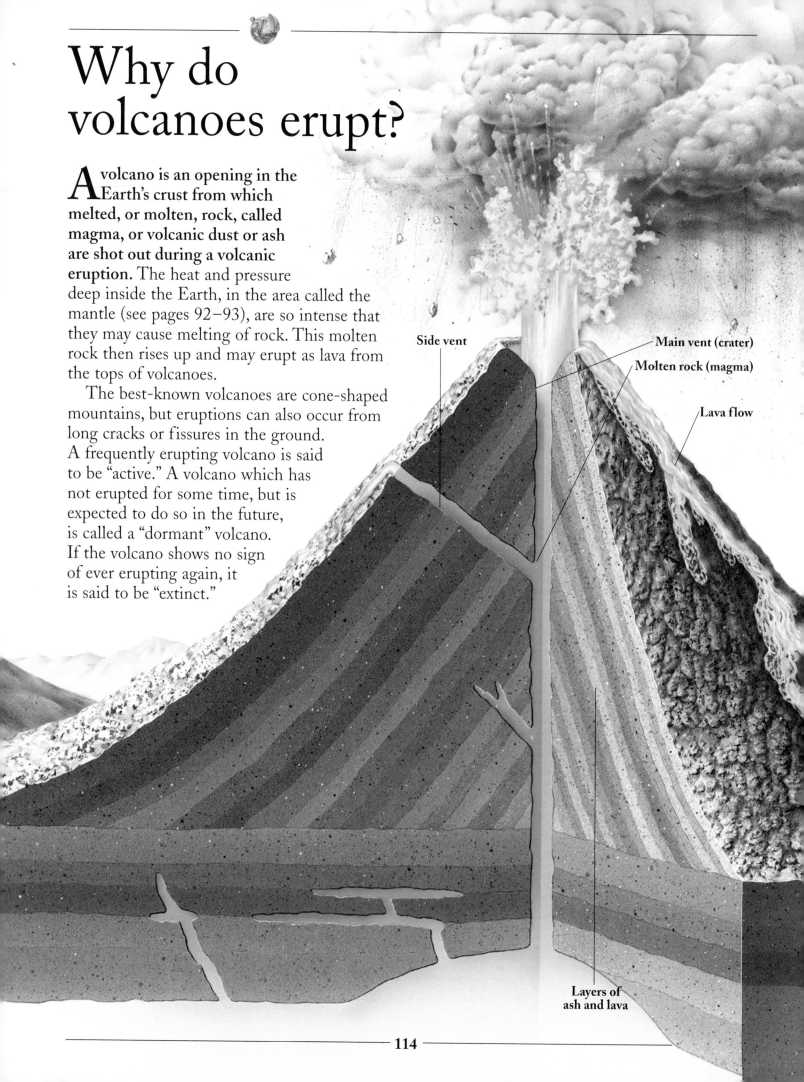

Why do volcanoes erupt?

A volcano is an opening in the Earth's crust from which melted, or molten, rock, called magma, or volcanic dust or ash are shot out during a volcanic eruption. The heat and pressure deep inside the Earth, in the area called the mantle (see pages 92–93), are so intense that they may cause melting of rock. This molten rock then rises up and may erupt as lava from the tops of volcanoes.

The best-known volcanoes are cone-shaped mountains, but eruptions can also occur from long cracks or fissures in the ground. A frequently erupting volcano is said to be "active." A volcano which has not erupted for some time, but is expected to do so in the future, is called a "dormant" volcano. If the volcano shows no sign of ever erupting again, it is said to be "extinct."

Side vent

Main vent (crater)

Molten rock (magma)

Lava flow

Layers of ash and lava

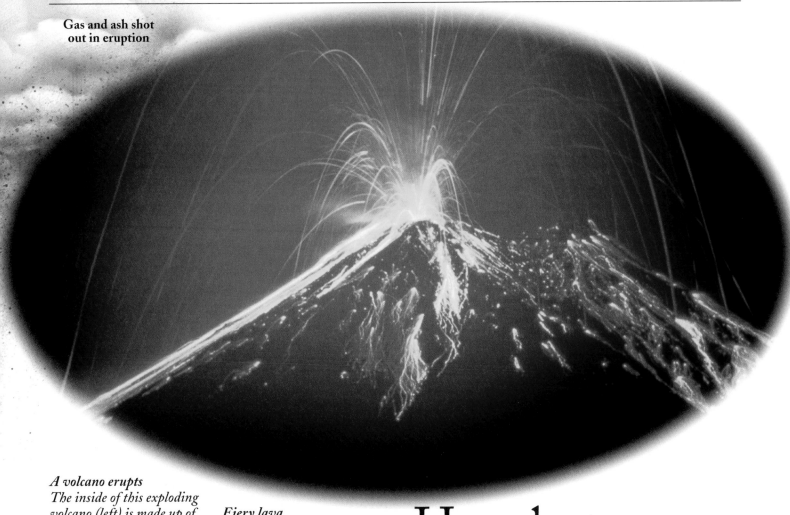

Gas and ash shot out in eruption

A volcano erupts
The inside of this exploding volcano (left) is made up of alternating layers of ash and lava from past eruptions.

Fiery lava
Red-hot lava pours from the exploding Arenal Volcano in Costa Rica during an eruption in July 1991.

How hot is it inside a volcano?

It must be hot enough inside a volcano to melt rock, because the lava that comes out of a volcano is liquid, or molten, rock. Inside the spaces and tunnels through which molten rock moves to the surface, the heat is incredibly intense. Most red-hot lavas emerge from a volcano at a temperature of between 1,650 and 2,200°F. Lava can flow rapidly out of the crater and can travel considerable distances into the surrounding area.

Deep down below the Earth's cool, hard crust is a layer called the mantle, where most magma comes from. This layer is like a giant cauldron, fired by the heat deep inside the Earth. Temperatures reach 7,800°F at the innermost layer, the Earth's core, and become cooler nearer the surface.

Are there different kinds of volcanoes?

Every volcano is different, but they can be grouped into several types, which have similar eruptions. A single eruption may produce a cinder cone of ash and fragments of lava. Later on in the eruption, lava may flow out through the base of the cone and one side of the cinder cone may collapse. Continued eruption of this type over many centuries at the same site may build up a composite volcano (see pages 114–115).

If a volcano produces flowing lava continually for months or even years, the flows spread far and wide from the vent. Brief individual eruptions give flows which heap one layer of lava on top of another. Longer eruptions of runny lava make shield volcanoes, which have a low, broad dome shape and can be miles across, such as the island of Hawaii.

Fissure volcanoes are caused by eruptions that occur along cracks in the land's surface. They also send out runny lava, but they are nearly flat in shape.

Volcano crater
A helicopter hovers over a huge mass of lava covering a vast stretch of land on Hawaii.

VOLCANOES

Cinder cone volcanoes are built up from erupted ash and lava fragments. They have the classic cone shape and a single vent.

Cinder cone volcano

Fissure volcanoes happen along faults and cracks in the Earth's surface. Lava emerges from many places along a particular line.

Fissure volcano

Shield volcanoes are low and flat, with many vents. They are formed from eruptions of runny lava over a long period.

Shield volcano

What is lava?

Lava is the red-hot molten rock that runs down the sides of an active volcano in great streams. If the lava is runny, it can move for great distances—perhaps tens of miles—before it finally stiffens and stops. The hole left at the top of the volcano is called the vent.

When the lava is less runny, explosive eruptions may happen. Lava fragments and volcanic ash are thrown over the countryside, and an ash layer several feet thick may blanket the land and form a new ground level. The ash is a wonderful fertilizer, and the ash and lava fragments are rapidly weathered and broken down to produce fertile soil—soil where crops can be grown.

The island of Surtsey near Iceland was formed by a series of powerful volcanic eruptions.

November 1963
A fissure opens in the seabed and liquid magma from the Earth's core is forced upward. A series of explosions hurls dust skyward.

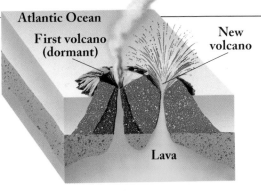

Atlantic Ocean
First volcano (dormant)
New volcano
Lava

February 1964
Near to the first cone, by this time dormant, a second volcanic peak, Surtur 2, begins to deposit lava on the island.

April 1964
Continued eruptions pour out more lava, making the island bigger. Surtsey today stands 570 feet above sea level at its highest point. There have been no eruptions since 1967.

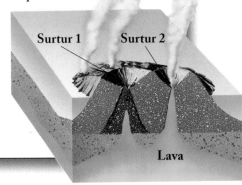

Surtur 1   Surtur 2
Lava

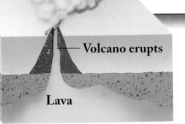

Atlantic Ocean
Volcano erupts
Lava

Buried in lava
A bus lies trapped in cooling lava at Kalapana, Hawaii. The village of Kalapana was buried under lava from the nearby Kilauea crater.

Can a volcano erupt under water?

Yes—in fact, most volcanic eruptions happen under water. Many deep underwater volcanoes erupt without anyone realizing. But a violent submarine volcanic eruption in shallow water can completely destroy an island. In 1883 one of the most powerful eruptions ever known on Earth tore apart the island of Krakatoa in the Pacific Ocean (see pages 106–107).

Volcanic eruptions can also create islands. In 1963, the black cone of a new volcanic island slowly built up out of the ocean off the coast of Iceland, as ash and then later bright red lava poured out from vents. Within a few weeks, the island stood more than 500 feet high and over a mile long. The island is called Surtsey after a legendary Norse giant.

Three-quarters of the Earth's surface—that is, all of the ocean floor—has been formed by deep eruptions under the sea. This happens in two ways. In some cases, lava may emerge from cracks and splits in the Earth's crust at ridges under the ocean. Lava can also come from single, "central vent" volcanoes that erupt on the ocean floor over "hotspots," where magma from deep inside the Earth (see pages 92–93) rises up through the crust.

Which are the world's biggest lakes?

The biggest lake in the world is the saltwater Caspian Sea in Asia. It has an area of 143,240 square miles. Lake Superior in North America is the largest of all freshwater lakes, with an area of 31,700 square miles. Lake Superior is the biggest of the five Great Lakes on the border of the United States and Canada.

The deepest of all lakes is Lake Baikal in Siberia, Russia. It is 5,712 feet deep, more than four times the depth of Lake Superior. It holds more fresh water than any other lake in the world—more than the five Great Lakes combined. Lake Baikal is also the oldest of all lakes and was probably formed about 25 million years ago.

What is a geyser?

A geyser is a natural hot-water fountain. The word geyser comes from the Icelandic name "geysir," which means "gusher." At a geyser, a scalding hot plume of boiling water, steam, and smelly gases shoots out of a hole in the ground and spouts upward for a matter of hours or even days.

Geysers happen in parts of the world where there are many active volcanoes, such as New Zealand, Iceland, and the northwestern United States. A geyser starts to form when water seeps down through crevices and cracks in rocks to fill an underground reservoir.

If this reservoir is positioned over rocks which are heated from below by magma, the molten rock inside the Earth, some of the water is turned to steam. Like a giant underground pressure cooker, this incredible heat produces a great surge in pressure, making water and steam shoot upward as a geyser fountain.

The world's tallest geyser spout was probably the one at Waimangu on the North Island of New Zealand. Its tower of water reached a height of about 1,500 feet. Old Faithful in Yellowstone National Park in Wyoming spouts to a height of 130 feet at regular intervals.

What is an iceberg?

Icebergs are huge masses of ice which break off from ice sheets or glaciers and float on the sea. They may be made of frozen seawater or fresh water and are found in polar areas. An iceberg eventually melts away and disappears. This melting can take a long time, however, if the iceberg is large and the sea around it is very cold.

Icebergs can be huge. One found floating near Antarctica in 1956 was larger than the country of Belgium. It covered an area of about 12,000 square miles.

What causes landslides and avalanches?

A landslide happens when soil, stones, and mud suddenly slip down a steep hillside or cliff. Excess water from heavy rainstorms or flooding makes soil so wet and heavy that it starts to move under its own weight. Once it starts moving, the mixture, containing stones and other debris, can flow almost like a thick river of sludge. Violent shaking can also make land slip, and landslides often happen as a direct result of earthquakes.

An avalanche is similar to a landslide, but it is large amounts of snow and ice that tumble down a snow-covered mountainside. Snow and ice avalanches are often triggered by a change in weather conditions. A period of heavy snowfall can build up thick, unstable layers of snow on the mountain slopes. If the weather suddenly gets warmer, some of this snow may melt, and a layer of water forms between the rock and the snow. The snow can then easily slide on this weak, wet layer, causing an avalanche.

What is a fjord?

Fjord is the Scandinavian word for a long, more or less straight, steep-sided valley. A fjord is filled by the sea and stretches inland for a considerable distance. These fjord valleys were cut by glaciers during past ice ages. When the glaciers melted at the end of the last cold phase of the current Ice Age, sea levels rose and the valleys were "drowned" with seawater.

Norway and New Zealand are famous for their many fjords. The Milford Sound Fjord in New Zealand is 12 miles long and is flanked by the world's tallest sea cliffs.

Are there rivers under the ground?

Yes. We usually think of rivers as flowing above the ground, but for parts of their courses, many rivers flow below the ground. Some rivers even start under the ground. If the highlands where a river begins are made of porous rock which soaks up moisture like a sponge, the ground absorbs lots of rainwater. The water then sinks into the ground instead of flowing on its surface. When it meets rock that is not so absorbent, the water changes course to flow on top of the new layer and comes to the surface as a spring—a gushing of water from under the ground.

In limestone areas, where slightly acidic rainwater dissolves the rock easily, large underground rivers are common. They may flow for miles under hillsides before coming out above the ground.

What is a mineral?

Minerals are the natural substances from which rocks are made. They are usually made of crystals, and each type has a particular chemical composition. A rock contains at least two different minerals.

The most common elements in rock-forming minerals are oxygen and silicon—oxygen is contained in oxides and silicon in silicates. The most common mineral is quartz, which is made of silicon dioxide.

What makes a stone precious?

A precious stone is one that is beautiful, rare, and valuable. Most gemstones are precious because their beauty and color make people want them, and their rarity makes them valuable. There are many different gemstones. Some of the best known are diamond, emerald, ruby, sapphire, garnet, and topaz. Most gemstones are found as natural crystals in rocks at the Earth's surface. They are then cut and polished to make them look as beautiful and as colorful as possible.

Gemstones also have industrial and scientific uses. Diamonds are the hardest natural mineral on Earth and are used on the tips of drills.

What is the atmosphere?

The atmosphere is the mixture of gases surrounding the Earth, the gases that we call "air." All over the planet, the surfaces of both land and sea are blanketed by this layer of gas, held in place by the force of gravity. About 99 percent of the atmosphere is made up of nitrogen and oxygen.

The atmosphere is made up of several layers. The layer closest to Earth is the troposphere, which extends about 10 miles above the Earth's surface. This contains most of the atmosphere's moisture—the clouds, rain, and snow. Next is the stratosphere, which is dry and contains the special form of oxygen called ozone. Ozone protects life on Earth by absorbing most of the Sun's ultraviolet rays.

From the stratosphere and mesosphere upward, the density of gases is too low for us to survive. Each breath would contain so little oxygen that we would suffocate. The temperature at the top of the mesophere is about −130°F, the coldest in the atmosphere. It is sometimes marked by special silver, rippling clouds called noctilucent clouds. Above 50 miles is the intensely hot thermosphere. This is followed at 300 miles by the exosphere which merges into outer space.

Where did the atmosphere come from?

The atmosphere came into existence soon after our planet formed, over four billion years ago. It slowly built up from the gases produced during volcanic eruptions and other phenomena. But that original layer of gases was quite unlike today's.

Scientists think that the early atmosphere was made up mainly of water vapor, hydrogen, carbon dioxide, and

Exosphere

Thermosphere

Electricity-charged layers

Space shuttle during re-entry

Atmosphere
The Earth's atmosphere is essential for our survival. It provides life-giving oxygen and protects us from harmful radiation.

Mesosphere

Ozone layer

Stratosphere

Troposphere

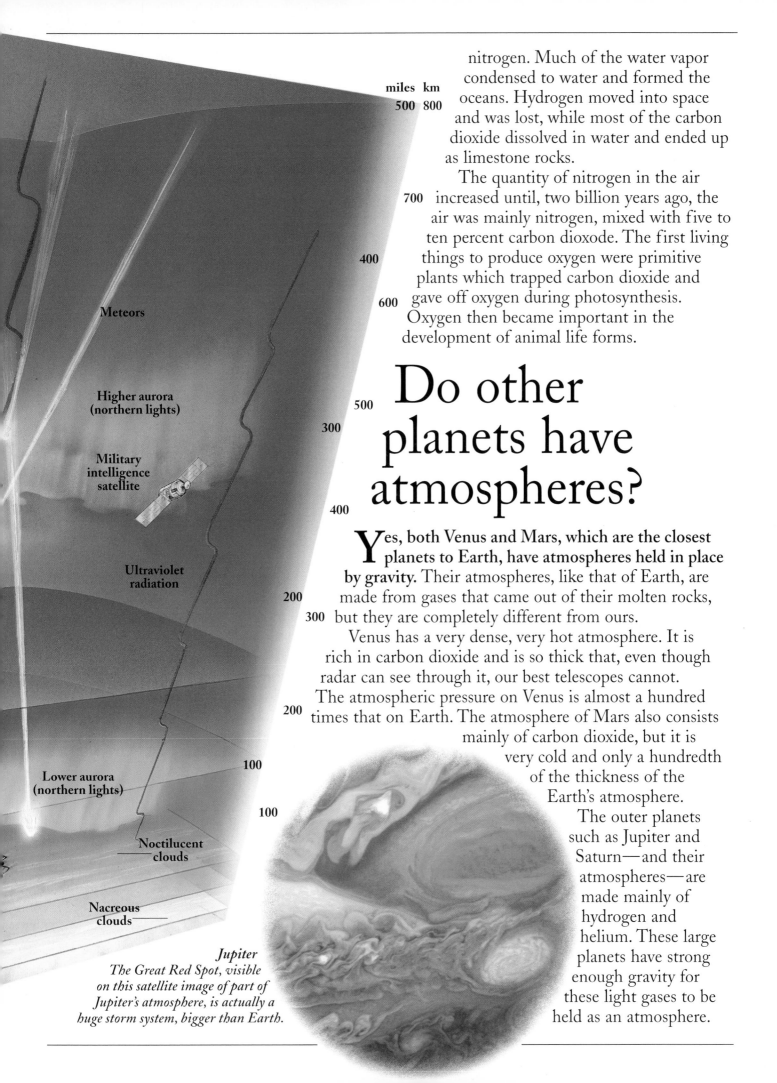

nitrogen. Much of the water vapor condensed to water and formed the oceans. Hydrogen moved into space and was lost, while most of the carbon dioxide dissolved in water and ended up as limestone rocks.

The quantity of nitrogen in the air increased until, two billion years ago, the air was mainly nitrogen, mixed with five to ten percent carbon dioxode. The first living things to produce oxygen were primitive plants which trapped carbon dioxide and gave off oxygen during photosynthesis. Oxygen then became important in the development of animal life forms.

Do other planets have atmospheres?

Yes, both Venus and Mars, which are the closest planets to Earth, have atmospheres held in place by gravity. Their atmospheres, like that of Earth, are made from gases that came out of their molten rocks, but they are completely different from ours.

Venus has a very dense, very hot atmosphere. It is rich in carbon dioxide and is so thick that, even though radar can see through it, our best telescopes cannot. The atmospheric pressure on Venus is almost a hundred times that on Earth. The atmosphere of Mars also consists mainly of carbon dioxide, but it is very cold and only a hundredth of the thickness of the Earth's atmosphere.

The outer planets such as Jupiter and Saturn—and their atmospheres—are made mainly of hydrogen and helium. These large planets have strong enough gravity for these light gases to be held as an atmosphere.

Jupiter
The Great Red Spot, visible on this satellite image of part of Jupiter's atmosphere, is actually a huge storm system, bigger than Earth.

miles km
500 800

700

400

600

Meteors

500

300

Higher aurora (northern lights)

Military intelligence satellite

400

Ultraviolet radiation

200

300

200

100

Lower aurora (northern lights)

100

Noctilucent clouds

Nacreous clouds

How is it possible to predict the weather?

It is possible to predict the weather of the future because of weather records that have been kept in the past. For over a hundred years, weather scientists, called meteorologists, have carefully observed and studied the main patterns of weather on most parts of the globe. They have mapped out their observations and records, and these help them tell us what the general weather patterns are, though there may be weekly or daily variations.

The Earth's weather systems work like an enormous machine. The fuel that powers the machine is heat energy coming to us across 90 million miles of space from the Sun. That energy drives all the changes and movements in the atmosphere which, together, we call weather.

What does a weather satellite do?

A weather satellite's job is to send radio signals about the atmosphere back to national and international weather centers, 24 hours a day. This information can then be used to help make detailed, accurate, continually updated weather forecasts.

Weather satellites are launched into space by rockets and then left to orbit the Earth.

Polar satellite orbit

Earth's spin

Polar satellites
Some satellites (left) orbit about 500 miles up and cross over the poles. They can take very detailed pictures of the strip of land over which they pass and send them back to Earth.

They are fitted with very sensitive cameras, which send images coded into radio signals back to Earth. The cameras record the appearance of the clouds, and they also measure the heat radiating out from the atmosphere. These measurements tell scientists the temperature and moisture content of the air at different heights. Measuring the distance the cloud belts move between one picture and another gives information about wind speeds.

Geostationary satellites
These satellites (right) orbit about 22,300 miles up above the Earth's surface and can take pictures of an entire hemisphere at once. Their orbit takes 24 hours—the same as one spin of the Earth.

What do patterns on a weather map mean?

The special lines and symbols drawn on a weather map may be difficult for anyone but a meteorologist—a weather scientist—to understand, because they show the position of things that cannot be seen.

The main patterns are lines called isobars. These are a bit like contour lines on a geography map, but they join points with the same atmospheric pressure instead of points of equal height above sea level. Zones of high and low pressure are labeled as H or L, showing the centers of the high and low pressure zones.

The atmospheric pressure is constantly changing from one area to another on the Earth's surface. Most of the change is brought about by differences in the temperature and moisture content of the air. Although you cannot see differences in atmospheric pressure, their effects are very important. They determine, for instance, how fast and in what direction winds blow. Wind speed is a result of isobar "packing"—the closer the lines are together, the stronger the wind blows.

Other symbols on the weather maps show the edges of large masses of cold or warm air and the direction they are moving. These edges are called fronts. Cold fronts are usually drawn in blue and warm fronts in red. Detailed weather maps also have symbols that show the type and amount of cloud, the wind direction, the amount of moisture in the air (the humidity), and the level of visibility.

Computerized weather
Advanced technology such as satellites and computer programs have revolutionized weather forecasting. High-speed computers are needed to make sense of the vast amounts of weather information constantly arriving from the satellites, radar, and other weather stations, and analyze it for the forecasters.

Where does the wind come from?

Wind is the movement of a mass of air. This movement is caused by differences in pressure in the atmosphere. A wind always flows from a high-pressure area to a low-pressure area, trying to equal out the pressure. It is rather like the air in a balloon, which is at a higher pressure than the air outside. If a puncture happens, air rushes out of the balloon, from the high to the low area of pressure.

Over the equator is a band of low pressure, and over each of the poles is a band of high pressure. There are alternating bands of high and low pressure over the rest of the planet, and they control the most common wind directions. These patterns, and the effect of the Earth spinning on its axis, create winds which generally blow east and west, not north and south.

Winds and waves
Strong winds can cause much damage to buildings and the land. Winds can also whip up huge waves which crash against the shore.

Wind patterns
The map shows the main patterns of high and low pressure and wind movements on Earth.

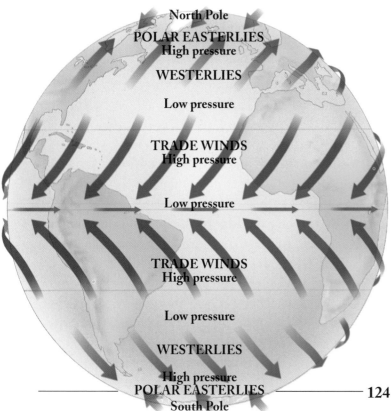

North Pole
POLAR EASTERLIES
High pressure
WESTERLIES
Low pressure
TRADE WINDS
High pressure
Low pressure
TRADE WINDS
High pressure
Low pressure
WESTERLIES
High pressure
POLAR EASTERLIES
South Pole

What is a hurricane?

When the wind speed of a tropical storm is more than 74 miles an hour, the storm is called a hurricane. These vast swirling and spiraling weather systems may be over 1,000 miles across, with winds of up to 200 miles an hour. They also feature huge rainclouds that drop rain at an astounding rate. As much rain may fall on one spot in a single day during an intense hurricane storm as falls on London or Seattle in a year.

All these storms begin in the tropics. They are called hurricanes in the Atlantic, typhoons off the China coast, cyclones or tropical cyclones in the seas off India, and willy-willies in Australia.

Whatever its name, each storm begins as an area of intensely low pressure spiraling over the sea. When the sea temperature reaches 81°F or more, the spiral grows bigger and bigger. In an area of low pressure, warm, moist air is sucked up into the atmosphere. When this air cools, the water vapor in it changes to water (condenses). As it does so, staggering amounts of energy are released, and this energy fuels the great speed and fury of this tropical storm.

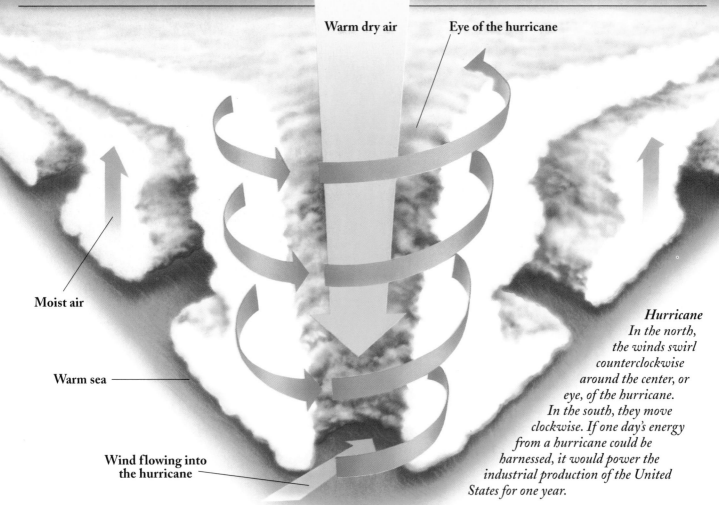

Warm dry air

Eye of the hurricane

Moist air

Warm sea

Wind flowing into
the hurricane

Hurricane
In the north,
the winds swirl
counterclockwise
around the center, or
eye, of the hurricane.
In the south, they move
clockwise. If one day's energy
from a hurricane could be
harnessed, it would power the
industrial production of the United
States for one year.

What causes a tornado?

A tornado is a terrifying funnel of rapidly rotating air. It drops from thunderstorm clouds and sucks up almost anything movable in its path. A tornado develops from a small but intense area of low pressure in the air below a storm cloud. If this pressure becomes powerful enough, it starts off a tight spiral of rushing air that is pulled in from all directions. If this funnel of air reaches the ground, it pulls in soil, debris, and anything that is not secure.

Over a small area, winds in a tornado may reach speeds of 370 miles an hour. The width of the funnel zone is usually only between 300 and 2,000 feet. Tornadoes generally last only 20 to 30 minutes, but one in Texas in 1917 lasted for seven hours. During their short lifetime, tornadoes travel across the countryside at the rate of 40 miles an hour or so. Though they are smaller than hurricanes, tornadoes can be more destructive.

Whirlwinds are less violent air spirals which are created when the Sun heats the ground and makes a whirling column of hot air lift off and spiral upward. Whirlwinds pick up sand, dry dusty soil, or even snow.

Tornado
The section of a tornado that is visible below the clouds
and reaches the ground is only part of the powerful
twisting structure. The spiraling air movements extend
right up into the cloud itself.

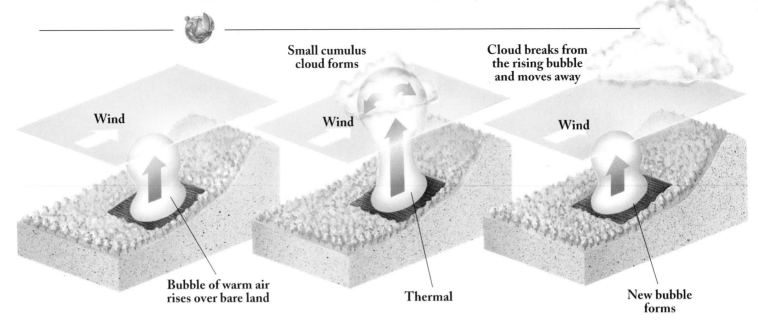

Small cumulus cloud forms

Cloud breaks from the rising bubble and moves away

Wind

Wind

Wind

Bubble of warm air rises over bare land

Thermal

New bubble forms

How a cloud forms
As warm air rises, it starts to cool off. Its water vapor turns into tiny drops of water, forming a cloud. The wind detaches the cloud from the rising air and it drifts away.

How do clouds form?

A cloud is a huge mass of air containing droplets of water or ice. All air contains water vapor, but when a mass of air cools down, the water vapor in it turns into tiny drops of water. Individually, these minute water droplets are invisible—each one is only about a millionth part of a normal raindrop. But together they make up the beautiful white mass of a cloud.

The temperature below which the vapor starts changing to water droplets is called its "dew point." This point is most often reached when a mass of moist air rises upward from the Earth—for example, as it moves over a mountain. As it moves upward in the atmosphere, it travels into areas of lower and lower air pressure, where it can easily expand. As the cloud gets bigger, it also gets cooler.

Up in the clouds
A layer of fluffy white cumulus clouds seen from an airplane flying above them. Cumulus clouds usually hover about a mile above the ground.

Are there different sorts of clouds?

Yes. Each type of cloud is given a different name, depending on its height above the ground, and its shape. Clouds are broadly divided into three groups according to the way that they are formed: cumulus (heaped clouds), stratus (layered clouds), and cirrus (feathery clouds). Somewhere around 35,000 feet above sea level—about 5,000 feet higher than the peak of Mount Everest—is a region called the tropopause. This region is the boundary between the dry, smooth, very thin air of the stratosphere, and the moist, stirring, thicker air below, called the troposphere (see pages 120–121). With few exceptions, clouds are found only in the troposphere. You realize this when you look out from the windows of an ascending airplane—as the plane gets higher in the sky, you find you are above the level of the clouds.

The very highest types of cloud are cirrus: thin, wispy formations, often found above 30,000 feet. But stratus and cumulus clouds usually form at 1,000 feet or less.

Why do some clouds make rain and others snow?

The difference between how rain and snow are produced seems to have as much to do with changes of temperature in the air as with the type of cloud concerned.

In most clouds, the water droplets are so tiny that they stay suspended in the air. They move up and down within the cloud, like the "sunbeam" dust particles we see dancing in the air lit by a ray of sunlight. If the cloud gets cold enough, the droplets will change to ice crystals. These are heavier than water droplets and begin to drift downward, under the force of gravity.

The top of a cloud is colder than the bottom and will reach the freezing point needed to make ice crystals first. If, by the time these crystals reach the bottom of the cloud, the temperature has risen enough for them to melt, the crystals change into large water drops, which fall as rain. But if the bottom of the cloud and the air below it are cold enough for the ice crystals to stay frozen, they drift down to Earth as snowflakes.

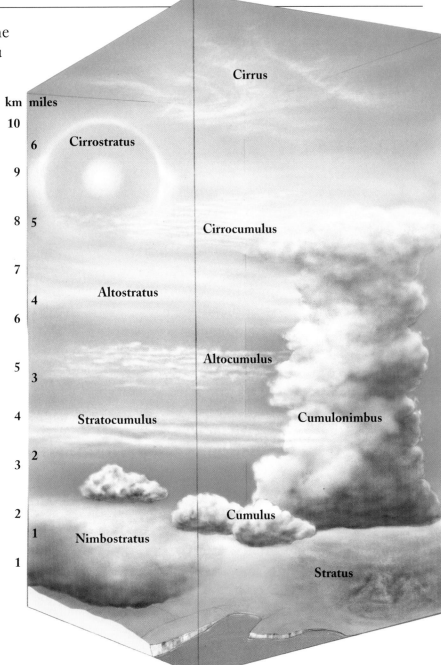

km miles
10
6   **Cirrostratus**
9
8   5
7
4   **Altostratus**
6
5   3
4   **Stratocumulus**
3   2
2   **Cumulus**
1   **Nimbostratus**
1

Cirrus

Cirrocumulus

Altocumulus

Cumulonimbus

Stratus

Cloud types
There are ten main types of cloud. They can usually be recognized by their shape and the height at which they form. Many are linked to particular weather. Cumulonimbus, for example, bring stormy showers of rain, hail, or snow.

SNOWFLAKES

Each large snowflake is made up of thousands of tiny six-sided crystals of many different patterns. The crystals may be feathery six-rayed stars or flat six-sided plates, depending on the temperature and moisture content of the cloud they come from. The way in which the tiny crystals join together varies from one flake to the next. This means that no two snowflakes are alike, even among the billions that fall in a flurry of snow.

Which is the coldest continent?

Antarctica is not only the coldest continent but also the windiest and driest place in the world. It has an area of more than 5 million square miles and is the fifth largest continent, bigger than Europe or Australia.

The continent is almost completely covered by a layer of ice which is nearly 2 miles thick in winter. Temperatures in winter can fall as low as −100°F, and the average for the year at the South Pole is only −58°F. Only on the Antarctic Peninsula, the piece of land which juts out across the Antarctic Circle toward South America, does the temperature creep above freezing in summer.

The highest point on Antarctica is the Vinson Massif in the Ellsworth Mountains, which is 16,066 feet high. There is also an active volcano called Mount Erebus.

Unlike other continents, Antarctica is not divided into different countries, and nobody lives there permanently. In the 19th century, people catching seals and whales had bases on Antarctica. Now scientists work there in research stations, most of which are on the Antarctic Peninsula. For now at least, this barren but beautiful land remains one of Earth's great wildernesses.

Which is the biggest continent?

Asia is the largest continent. It has an area of more than 17 million square miles and covers nearly a third of the world's land. It includes the world's highest point—Mount Everest, which is 29,028 feet high—and one of the lowest—the Dead Sea. Asia extends through many different climates, from Siberia in the far north, some of which is inside the Arctic Circle, to the tropical islands of Southeast Asia on the equator.

The Chiang Jiang is the longest river in Asia and one of the longest in the world. It is about 3,964 miles long. The largest saltwater lake in the world, the Caspian Sea, is also in Asia.

More people live in Asia than in any other continent. The population is more than 3 billion. China alone has a population of more than one billion, the most people of any country in the world. There are more than forty different countries in Asia.

Is Europe a continent?

Europe is the sixth largest continent in the world with an area of more than 4 million square miles. The boundary between the continents of Europe and Asia is generally thought to be the Ural Mountains.

Although Europe has only about 7 percent of world's land area, it has the second largest population after Asia. More than 690 million people live in Europe.

The highest points in Europe are Elbrus in Russia, which is 18,510 feet high, and Mont Blanc in the French-Italian Alps, which is 15,770 feet high. The longest river is the Volga, which is 2,290 miles long.

Is Africa the hottest continent?

Most of Africa is within the tropics, and it is the hottest of all the continents. The world's highest-ever temperature,

136.4°F in the shade, was recorded in Libya in North Africa. Africa is the second largest continent and covers an area of 11,687,183 square miles. Its highest point is Mount Kilimanjaro at 19,340 feet. Africa has the biggest hot desert in the world, the Sahara, and the world's longest river, the Nile. The Nile is about 4,145 miles long.

South of the Sahara around the equator is an area of tropical rain forest. To the east and south of this are the great savannas where animals such as zebras, wildebeest, and lions roam. In the far south are more deserts—the Kalahari and the Namib.

Is North America a separate continent?

North and South America are separate continents. North America includes the narrow piece of land called Central America to where the country of Panama borders Colombia in South America.

North America is the third largest of the continents with an area of 6,880,635 square miles. It has a population of more than 415 million. Its landscape is extremely varied, ranging from the northern lands within the Arctic Circle to the tropical forests of Central America. The west is dominated by the Rocky Mountains, which run from Alaska to Mexico. Its highest point, and the tallest mountain in North America, is Mount McKinley which is 20,320 feet.

The world's largest freshwater lake, Lake Superior, is in North America, and together with the other Great Lakes makes up the world's largest body of fresh water. Off the coast of Canada is Greenland, the world's largest island. The longest river is the Mississippi-Missouri at 3,710 miles long.

Is South America smaller than North America?

South America is smaller than North America. It is the fourth largest continent and has an area of 6,886,000 square miles. A land of contrasts, South America extends from the tropical forests in the north to the bleak Patagonian Desert in the far south. The world's longest mountain chain, the Andes, stretches all the way down the western side of the continent. The highest point is Aconcagua at 22,834 feet.

The Amazon River in South America is second to the Nile of Africa in length but carries more water than any other river. It flows through the great Amazon rain forest, the largest in the world and home to an extraordinary variety of animals and plants.

Which is the smallest continent?

Australasia is the smallest of all the continents, with an area of 3,445,197 square miles. It includes not only Australia but also New Zealand, Papua New Guinea, and thousands of Pacific islands. Only about 26 million people live in Australasia, fewer than in any other continent except Antarctica.

The highest point in Australasia is Mount Wilhelm in Papua New Guinea, which is 14,793 feet. The highest point in Australia itself is Mount Kosciusko, which is 7,310 feet.

What makes lightning happen?

A **lightning flash is a huge electrical spark in the sky, caused by the immense charges of static electricity that can build up in thunderclouds.** The spark usually jumps between the bottom of a thundercloud and the ground beneath it, then back up again, as opposite charges cancel each other out.

Thunderstorms develop from large cumulonimbus clouds (see pages 126–127), which reach a long way up into the sky. Very powerful currents of air move up and down repeatedly in these clouds. They are the same currents that can produce hailstones. But the movements of air and ice can also cause great electrical charges in the clouds.

There are usually strong negative charges near the bottom of the cloud, and positive ones in the middle of the cloud. When the lower regions of the cloud pass over the land beneath, they bring about an opposite positive charge on the ground beneath them. A flash of lightning happens when these charges get so enormous that the usual electrical resistance of the air collapses.

What is the difference between sheet and fork lightning?

T **here is really no difference between sheet and fork lightning.** Both are electrical sparks caused by static electrical discharges. Fork lightning is the typical jagged stroke that can be clearly seen between the cloud and the ground. Sheet lightning is simply a stroke of lightning hidden from clear view by intervening clouds, so it looks like the sky is just briefly lit up.

Lightning usually starts as a "leader stroke," the main lightning stroke which runs swiftly from the bottom of the cloud—bristling with negative static electricity—to the highest point on the ground below. There are then one or more return strokes, which flash from the ground back up into the cloud, producing the "forked" or jagged effect that lights up the darkened sky.

The highest point on the ground may be a tree, or the spire of a church, and this high object may be "struck" by lightning. A person standing under a tree during a lightning strike can be killed by the powerful electrical force.

Cloud to air lightning

Flashes in the sky
A split second is caught on camera as flashes of lightning light up the night sky in Arizona during an electrical storm.

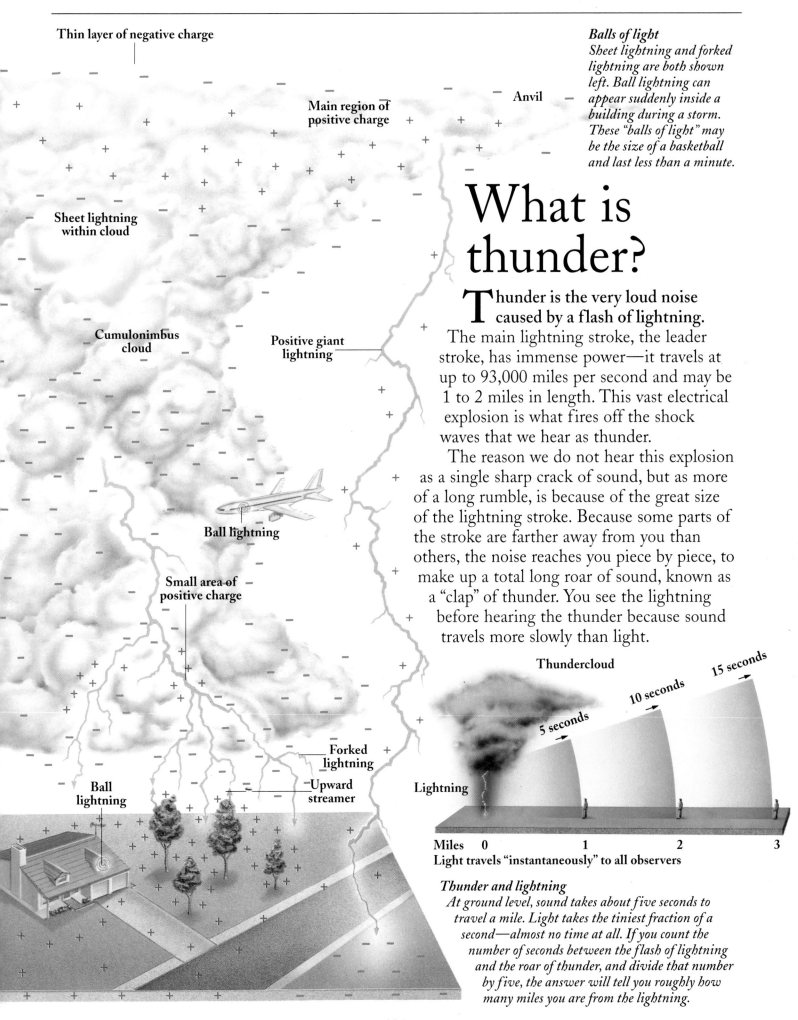

Thin layer of negative charge

Main region of positive charge

Anvil

Balls of light
Sheet lightning and forked lightning are both shown left. Ball lightning can appear suddenly inside a building during a storm. These "balls of light" may be the size of a basketball and last less than a minute.

Sheet lightning within cloud

What is thunder?

Cumulonimbus cloud

Positive giant lightning

Ball lightning

Small area of positive charge

Ball lightning

Forked lightning

Upward streamer

Thunder is the very loud noise caused by a flash of lightning. The main lightning stroke, the leader stroke, has immense power—it travels at up to 93,000 miles per second and may be 1 to 2 miles in length. This vast electrical explosion is what fires off the shock waves that we hear as thunder.

The reason we do not hear this explosion as a single sharp crack of sound, but as more of a long rumble, is because of the great size of the lightning stroke. Because some parts of the stroke are farther away from you than others, the noise reaches you piece by piece, to make up a total long roar of sound, known as a "clap" of thunder. You see the lightning before hearing the thunder because sound travels more slowly than light.

Thundercloud

15 seconds

10 seconds

5 seconds

Lightning

Miles   0   1   2   3

Light travels "instantaneously" to all observers

Thunder and lightning
At ground level, sound takes about five seconds to travel a mile. Light takes the tiniest fraction of a second—almost no time at all. If you count the number of seconds between the flash of lightning and the roar of thunder, and divide that number by five, the answer will tell you roughly how many miles you are from the lightning.

Why does a rainbow have so many colors?

A rainbow has seven colors: red, orange, yellow, green, blue, indigo (dark blue), and violet, always in the same order. These are the colors of the spectrum from which all light is made up. Normally they are blended together and so are invisible.

Raindrops can split up sunlight in the same way, producing a kind of spectrum in the sky, which we call a rainbow. As the Sun's rays enter a droplet of water, they are bent and split up into the colors of the rainbow. This rainbow within the raindrops is reflected back from the far surface and is bent again as it emerges from the raindrops. It is this bending which causes the distinctive arc of a rainbow.

Rainbows caused by light passing through the water of waterfalls or fountains may last for hours rather than minutes—as can be seen at Niagara Falls, for example.

Light reflected by raindrop

Sunlight

42°

Rainbow

Observer's line of vision

Spectrum

HOW A RAINBOW FORMS

Sir Isaac Newton, the 17th-century English scientist, showed that white light is really a mixture when he passed ordinary sunlight—white light—through an angled lump of glass. The white light was split by its passage through the glass into the range of colors of which it is made up. For a rainbow to appear in the sky, there must be rain and sunshine at the same time. A rainbow is made when Sun shines on raindrops. The rays of sunshine are bent by droplets of water, which break them up into their seven colors. Each drop of rain reflects a spot of colored light, and millions of these spots join together to form a rainbow.

Rainbow
The larger the raindrops, the brighter the colors of a rainbow. Here, a glorious rainbow has formed over mountains in Arizona.

Where is the rainbow's end?

Nowhere that you could **ever find it!** You can no more reach the end of a rainbow than you can touch your reflection in a mirror. The rainbow and the reflection are both what scientists call "virtual" images. They are caused by light reaching your eyes in a way which suggests that it is coming from a specific distant point, when in fact it is not. There is not really another object like your head behind the mirror, nor is there a brightly glowing arc at a certain place in the sky when you glimpse a rainbow shimmering in the sky.

Rainbows have always been thought of as miraculous—and they have inspired many legends and fairy tales. The ancient Greeks, for example, imagined that the rainbow was Iris, the messenger of the gods. Part of the reason for all these stories is that nobody really understood how rainbows were made. Now that we understand more about light and its effects, the rainbow is less of a mystery, but its beauty and its short life are still amazing.

One of the most popular stories about rainbows is that there is pot of gold to be found at its end. But of course, nobody has ever reached the end of a rainbow to claim the prize—or to prove or disprove the story!

What are the northern lights?

The northern lights are patterns of moving multicolored light seen in the night skies of the northern regions of the world. Similar southern lights are seen in Antarctica.

The lights are caused indirectly by the Sun. The outer atmosphere of the Sun gives off a solar wind, which carries highly energetic particles. Occasionally, solar flares burst out from the Sun, bringing sudden increases in the wind and giving off more particles.

The particles are attracted to Earth's magnetism and rain down toward it. The magnetism is strongest near the North and South poles, and here the solar wind and particles meet the gases in our atmosphere and make them glow.

Light show
The spectacular northern lights (below) are also called the aurora borealis. The similar colored lights seen in skies around the South Pole are called the aurora australis.

What causes an eclipse of the Sun?

An eclipse of the Sun happens when the Moon comes between the Earth and the Sun. It is called a partial solar eclipse if the Sun is only partly hidden, and a total eclipse if the Moon completely blocks out the Sun. A total solar eclipse may last for seven and a half minutes, but usually lasts only two to three minutes. During this time the landscape darkens dramatically, and birds and animals act as if night is approaching.

Although the Moon and Sun are vastly different both in size and in distance from us, they appear to be almost exactly the same size when viewed from Earth. This makes it possible for the Moon to cover the Sun's disk totally and block out its light.

The Moon's orbit around the Earth does not exactly correspond with the track that the Sun seems to take across the sky. If it did, there would be a total eclipse of the Sun every month. The two tracks are at an angle of about 5 degrees away from each other, which makes eclipses of the Sun rare events. Even so, there are between two and five solar eclipses every year, visible in some part of the world.

The solar eclipse of July, 1991, taken from Hawaii.

Are lunar eclipses caused in the same way?

Not quite. A solar eclipse happens when the Moon blocks the Sun from our view. An eclipse of the Moon takes place when the Earth's shadow, cast by the Sun, covers up part or all of the Moon. This only happens when the Sun and the Moon are on opposite sides of the sky as viewed from the Earth.

Another difference is that, while total eclipses of the Sun are only visible along a thin track across the Earth, lunar eclipses can be seen over much of the planet. This is because the Earth's shadow is positioned over the Moon, from wherever you observe it. The next total lunar eclipse is in September 1997. The next solar eclipses visible in the United States are in March 1997 and February 1998.

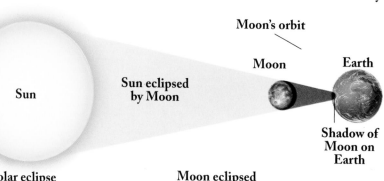

Moon's orbit

Moon **Earth**

Sun eclipsed by Moon

Sun

Shadow of Moon on Earth

Solar eclipse

Moon's orbit

Moon eclipsed by Earth's shadow

Sun

Earth **Moon**

Lunar eclipse

During a total eclipse of the Sun, the Moon passes between the Earth and the Sun. At the peak of the eclipse, the Sun disappears for a few minutes.

In an eclipse of the Moon, the Moon's face is slowly covered by the Earth's shadow.

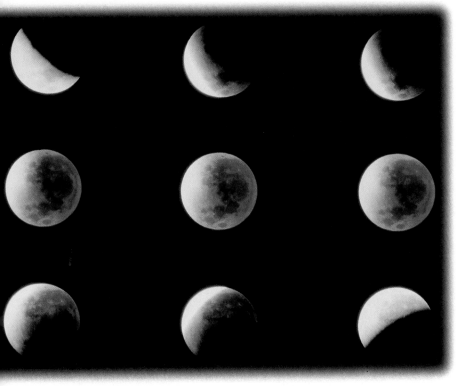

Eclipse of the Moon
Nine stages in the lunar eclipse of November 1993.

Why does the Moon appear to change shape?

The Moon does not change shape at all; what varies is the amount of its surface that we can see from Earth. The rest merges with the dark night sky and is invisible. The Moon is lit by light from the Sun. During the four weeks it takes the Moon to orbit the Earth, we see the lit and shadowed halves of the Moon from different angles.

The changes in the shape of the illuminated part of the Moon are known as "phases." At the beginning of the four-week cycle, a thin, bright crescent appears on the right-hand side of the Moon. It slowly gets bigger until all the right-hand side is bright. The lit portion then spreads farther to the left until the whole Moon's face is illuminated; this is a "full Moon."

The process then goes into reverse—a shadow appears to eat away a crescent shape on the right-hand side of the Moon. This shadow expands until the entire right-hand side is black, and continues until the whole disk is darkened. The Moon's sequence of "phases" then starts all over again.

Phases of the Moon
This sequence of photographs show some of the stages in the second half of the Moon's cycle as it wanes from full Moon to crescent.

Why is it hot in summer and cold in winter?

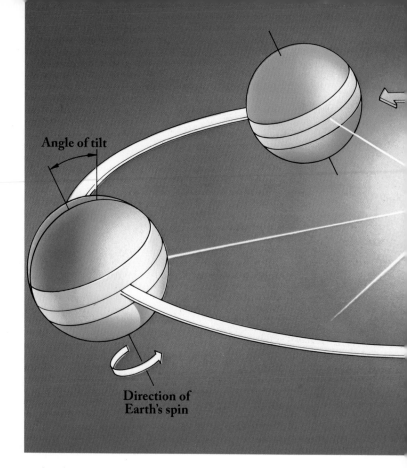

Angle of tilt

Direction of
Earth's spin

The contrast between the heat of summer and cold of winter is a result of the way in which the Earth tilts as it spins. Its axis, the imaginary line passing through the Earth from the North Pole to the South Pole, and around which it spins, is at an angle of about 23½ degrees. This means that the Earth spins at an angle to the path of the Sun's rays.

For half of its orbit around the Sun, and therefore for half the year, the northern half of the Earth is tilting toward the Sun. During the period of sunward tilt, the Sun is higher in the sky each day, and more direct light and heat reach the ground. This extra energy from the Sun makes the weather hotter in that half of the world.

The rest of the year, the northern half of the world is tilted away from the Sun. The Sun is lower in the sky each day, less energy from its rays get to Earth, and the weather is colder. This pattern is just the same in the southern half of the world, but in reverse. When the northern half of the world is tilted toward the Sun, the southern half is tilted away. So, when it is summer in Sydney, it is winter in New York and London.

The temperate areas, those between the tropics and the polar zones, have the biggest seasonal changes in temperature. In the tropics, the area around the equator, the tilt makes little difference—the weather is always hot. At the polar regions it is always cold, but it is much colder in the winter when the pole is tilted away from the Sun, and there is constant darkness for a time. In summer, the pole is tilted toward the Sun, and there is constant daylight for a time.

If the Earth did not tilt, the temperature for any place would be the same every day of the year.

Are there seasons in hot countries?

There are seasons in hot countries, but they are marked more by changes in the amount of rain that falls than by temperature differences.

In the tropical countries bordering the equator, there is often a main dry season and one or two wet seasons each year. All of the year's total of rain for the area will fall during these wet seasons, and plant life grows fast to take advantage of the water that is available.

Some parts of the tropics, though, have scarcely any seasonal change. Rain forests, such as those of the Amazon, are hot and wet all year, while deserts like the Sahara are always hot and dry.

The Sun's rays
In the tropics, the Sun's rays fall directly to the ground and bring intense heat. Farther north and south, the Sun's rays fall at an angle and spread over a large area, giving less heat overall.

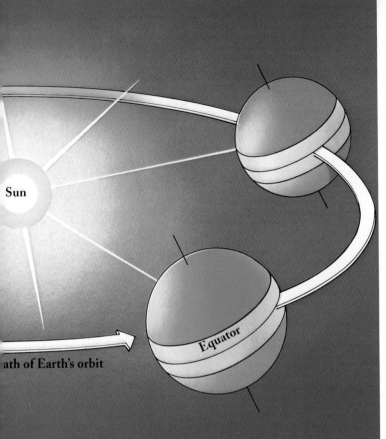

around the middle of the globe.) In the tropics, the Sun is always high in the sky, so its rays fall directly to the ground, and for at least one day a year it is directly overhead. Tropical days are about 12 hours of light, 12 of darkness, and do not vary much over the year. Such regular direct sunlight means a reliably hot climate all year round.

Near the North and South poles, even during the brief summer, the Sun is only just above the horizon, like midwinter Sun farther south. This low sunlight with its slanting weak rays cannot warm the ground very well.

Why is it hot in Africa and cold in the Arctic?

Africa is hot and the Arctic and Antarctic are always cold because of the way sunlight falls on these parts of the Earth.

A large area of Africa is in the tropics, the band of the Earth bordering each side of the equator. (The equator is an imaginary line

Hot and cold homes
In hot countries such as Thailand, houses can be light structures made of wood (above) and do not need to provide warmth. In the Arctic, igloos made of snow make surprisingly warm shelters (left).

Are we polluting the atmosphere?

We are polluting the atmosphere and with an increasingly wide range of polluting substances. There are so many people in the world now that we, and the industries we develop, produce too many pollutants to be absorbed by the atmosphere of our planet.

There are several important types of pollution that we introduce into the air. There are the acid gases that come from burning fuel such as coal and oil, which produce acid rain. This can cause serious damage to plants and animals. Burning almost anything at all produces the gas carbon dioxide. Increasing amounts of this gas pouring into the atmosphere create global warming called "the greenhouse effect." Gases from aerosol spray cans are another kind of airborne pollution.

Exhaust fumes from motor vehicles pollute the atmosphere by adding to the carbon dioxide. They are one of the causes of city smog. Where leaded gasoline is used, motor vehicles and airplanes also pollute the atmosphere and the ground with poisonous lead compounds.

Dangerous radioactive wastes in the atmosphere also used to contribute to pollution. But now that all atmospheric nuclear weapon testing has been stopped, this is much less of a problem. Some radioactive pollution comes from the rare accidents at nuclear power plants.

A traffic policeman in Bangkok wears a mask to protect himself against the ever-increasing levels of pollution in the city.

What is global warming?

Global warming is happening because human activities are pumping vast amounts of gases such as carbon dioxide, methane, and chlorofluorocarbons (CFCs) into the air. These gases are trapping more and more of the heat given off from the Earth's surface which normally escapes into outer space. This makes the atmosphere and the Earth's surface warmer—just like a greenhouse, whose glass walls and roof trap heat from the Sun.

Children born in the 21st century will probably grow up in a world much warmer than it is now. There could be shifting rainfall patterns, violent storms, and rising seas. There are already signs of a changing climate with average temperatures rising by 0.9°F in the 20th century. By 2050 the world could be hotter than it has been for 120,000 years.

If our planet's climate changes, there could be droughts in some areas and torrential rainfall and storm damage in other areas. Even worse, it could lead to the melting of parts of the polar ice caps. This could raise sea levels and flood vast areas around the edges of continents.

Water vapor

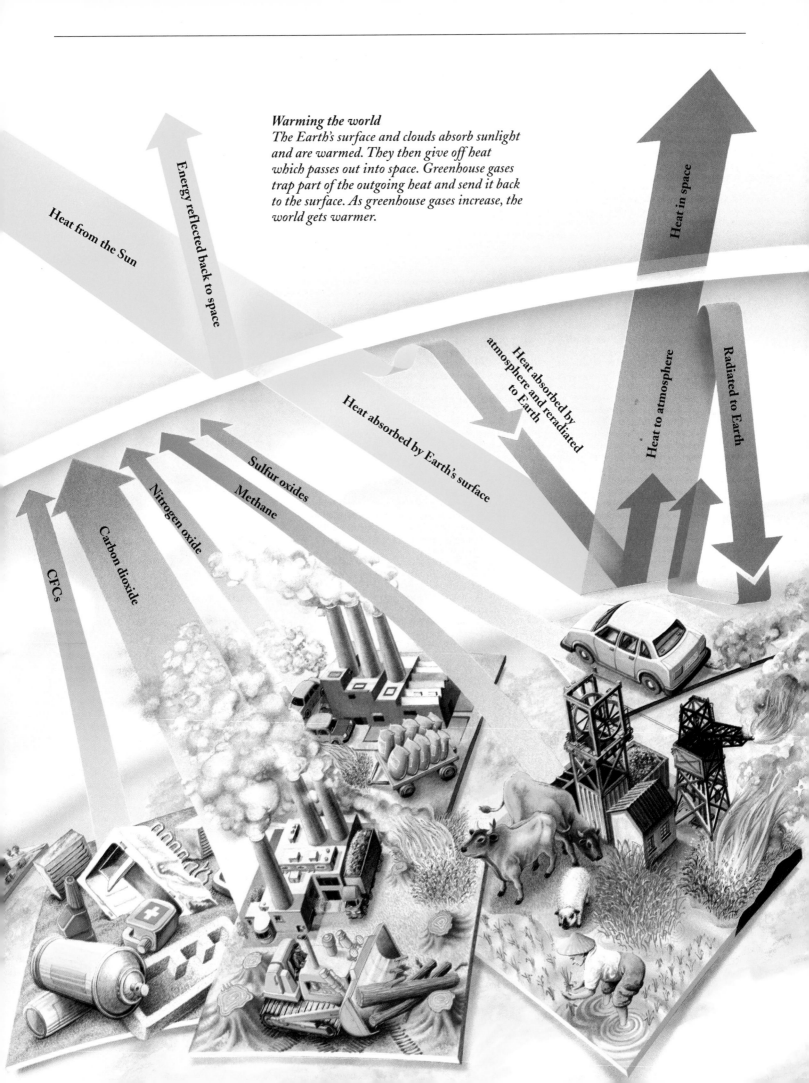

Warming the world
The Earth's surface and clouds absorb sunlight and are warmed. They then give off heat which passes out into space. Greenhouse gases trap part of the outgoing heat and send it back to the surface. As greenhouse gases increase, the world gets warmer.

Heat from the Sun

Energy reflected back to space

Heat in space

Heat absorbed by atmosphere and reradiated to Earth

Heat to atmosphere

Radiated to Earth

Heat absorbed by Earth's surface

CFCs

Carbon dioxide

Nitrogen oxide

Sulfur oxides

Methane

Glossary

You may find it helpful to know the meanings of some of these scientific words when reading the questions and answers in this book.

axis An imaginary line joining the North and South poles which runs through the center of the Earth. The Earth takes approximately 24 hours to make one complete turn on this axis.

bacteria Simple one-celled organisms which may live independent lives or depend for their existence on some other living creature. Many bacteria live in and on our bodies, and some of them cause diseases.

Carboniferous The period of time on Earth from 360 to 286 million years ago. The first reptiles lived in the Carboniferous period.

carnivore An animal that eats other animals in order to survive. Lions and tigers are examples of carnivores.

cell The basic unit of life. The simplest living plants and animals are made of one single cell. More complex organisms, such as mammals, contain billions of cells.

Cretaceous The period of time on Earth from 135 to 65 million years ago. Dinosaurs were at their most varied during this period.

deciduous Describes a tree or shrub that loses all its leaves at one season of the year.

electrical charge Within every atom there are tiny particles which attract or repel one another. These forces are known as electrical charges. There are two types of charge—positive and negative. Two positive or two negative charges will repel each other, while one positive and one negative attract each other. This is the basis of electricity.

erosion The wearing away and removal of material from the surface of the Earth. The pounding of ocean waves, for example, causes the erosion of the seashore. Water, wind, and ice are all powerful forces of erosion.

evaporation The process in which a liquid turns into a gas or vapor. When water is heated up, it evaporates into water vapor.

evergreen A plant (usually a tree or shrub) that does not lose all its leaves at one season of the year. Holly is an evergreen plant.

fault A break between two pieces of rock, or between plates of the Earth's crust. The presence of faults allows rocks to move.

fossil The remains of a plant or animal preserved in rock. Hard parts, such as teeth and bones, are more likely to form fossils than soft parts. Impressions in rock, such as footprints, can also become fossilized.

gravity The force of attraction between two bodies. It is gravity that

makes a ball fall straight down to the ground when it leaves your hand.

herbivore An animal that eats plants. Elephants and antelope are herbivores.

invertebrate An animal without a backbone. Insects and spiders, as well as creatures such as slugs, snails, crabs, shrimp, and clams, are invertebrates.

Jurassic The period of time on Earth from 195 to 135 million years ago. Dinosaurs, pterosaurs, and marine reptiles thrived at this time.

larva A young form of a creature, which hatches from an egg. A larva looks quite different from the adult animal. A caterpillar is an example of a larva.

magma Liquid, or molten, rock which is made deep inside the Earth. It can come to the surface as volcanic lava or be pushed up between the huge plates which make up the Earth's crust.

microbes Microscopic organisms, such as bacteria, that may cause disease.

mineral salts Chemicals found in fresh water, sea water, and in water in soil. Mineral salts are taken up by plants and used to help them survive and grow.

orbit The path taken by one object around another. The Earth orbits the Sun, while the Moon orbits the Earth.

parasite An animal or plant that depends on another living organism (its host) for food. Parasites may live on the outside or inside of their hosts' bodies.

photosynthesis The chemical reaction in which green plants trap energy from sunlight and use it to make foods, such as sugars, from the simple ingredients of water and the gas carbon dioxide.

plankton The tiny plants and animals suspended in the ocean and in fresh water.

pollen The tiny grains made by the male parts of a flower. Pollen must reach the female parts of a flower so that the plant can be fertilized and fruits and seeds can form.

pressure, atmospheric The force produced by the weight of all the air above any particular spot. The higher above the Earth you are, the lower the atmospheric pressure.

species A type of plant or animal. Living things of the same species can mate together and produce young which in turn are able to have young.

subtropics The parts of the world immediately north and south of the tropics.

temperate Regions of the world with climates that are not extremely hot or cold. Temperate areas have definite seasons—cool winters and warm summers.

Triassic The period of time on Earth from 245 to 195 million years ago. The first dinosaurs lived toward the end of this period.

tropics The hot regions of the world around the equator.

vertebrate An animal with a backbone. Fish, amphibians, reptiles, birds, and mammals are all vertebrates.

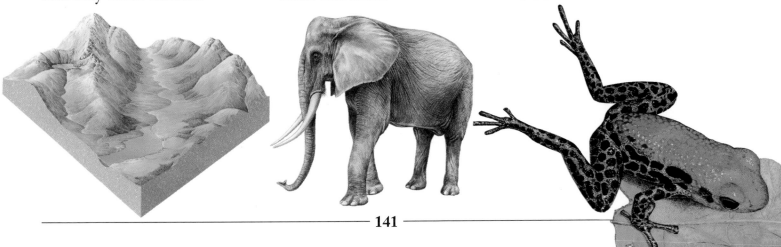

Index

Acknowledgments

Artwork

l = left; r = right; t = top; c = center; b = bottom

9l Andrew Wheatcroft; 9t and c Steve Kirk; 11 Joanne Cowne; 12–13 Colin Newman; 13t Mainline Design; 14t Roger Kent; 14–15 Steve Kirk; 16–17 Steve Kirk; 18–19 Elizabeth Gray; 20–21 Steve Kirk; 22–23 Steve Kirk; 24–25 Richard Phipps; 26t Eugene Fleury; 26–27 Andrew Wheatcroft; 29 Eugene Fleury; 30t Eugene Fleury; 30br Roger Kent; 31 (t to b) Colin Woolf, Dick Twinney, Michael Woods, Dick Twinney; 33tl Denys Ovenden; 33tr Colin Woolf; 33b Paul Richardson; 34–35 Richard Orr; 34t Colin Woolf; 35b Colin Woolf; 36 Joanne Cowne; 37tl Bridget James; 37tr and b Joanne Cowne; 38l Denys Ovenden; 38bl Norman Arlott; 38tr Hilary Burn; 38cr Chris Rose; 39r Michael Woods; 39b Norman Arlott; 40tl Peter Hayman; 40cl Michael Woods; 40bl Peter Hayman; 40t Michael Woods; 40cr Peter Hayman; 41 Norman Arlott; 42t and br Michael Woods; 42bl Peter Hayman; 42 maps Eugene Fleury; 43tl Graham Allen; 43tr Alan Male; 43b Dick Twinney; 44 Colin Newman; 45 Paul Richardson; 46t Eugene Fleury; 46 bl Eric Robson; 46br Gary Hincks; 48–49 Bernard Robinson; 50t Halli Verrinder; 50b Alan Male; 51 Eric Robson; 52tl and bl Joanne Cowne; 2cl and cr Elizabeth Key; 53c Alan Male; 53tr Michael Woods; 53br Eric Robson; 53bl Dick Twinney; 54t Eric Robson; 54bl Alan Male; 54–55 Denys Ovenden; 55t Colin Newman; 55tr Ken Wood; 55bc Owen Wood; 56l Joanne Cowne; 56r Dick Twinney; 57 Colin Woolf; 58tr Colin Woolf; 58bl Joanne Cowne; 58br Eric Robson; 58–59 Colin Newman; 59tl Alan Male; 59tr Colin Newman; 59br Graham Allen; 60–61 Simon Turvey; 62 Michael Woods; 63 Alan Male; 64t Julie Banyard; 64b Eric Robson; 64c Elizabeth Key; 66 David More; 67t David More; 67bl Elizabeth Key; 68 Eric Robson; 69t Graham Allen; 69c Chris Rose; 70 Tony Graham; 71t Eric Robson; 71c and 71br David More; 72 Julie Banyard; 73 Eugene Fleury; 74bl Alan Male; 74c Eric Robson; 75bl Graham Allen; 75tr Alan Male; 75c Michael Woods; 80c Graham Allen; 80–81 Eugene Fleury; 81t Colin Woolf; 81c Graham Allen; 81b Bernard Robinson; 82–83 Michael Woods; 82c Peter Hayman; 82b Michael Woods, Graham Allen; 83 Graham Allen; 84c Michael Woods; 84bl David Quinn; 84bc and br Graham Allen; 86t Alan Male; 86l Robert Gillmor; 86c and b Michael Woods; 87 Gary Hincks; 88t Graham Allen; 88b Michael Woods; 89tl and c Graham Allen; 89tc Ken Wood; 89tr Michael Woods; 91tr and tl Gary Hincks; 91bl Pavel Kostal; 92bl Andrew Farmer; 92–93 Mainline Design; 93tr Jonothan Potter; 94c Janos Marffy; 94–95 Pavel Kostal; 95t Gary Hincks; 96bl Gary Hincks; 96–97 Pavel Kostal; 97br Gary Hincks; 98–99 Pavel Kostal; 99tr Janos Marffy; 100–101 Mike Saunders; 103 Eugene Fleury; 104bl Gary Hincks; 104–105 Gary Hincks; 105 Steve Weston/Linden Artists; 107 Brian Watson/Linden Artists; 108 Eugene Fleury; 109 Janos Marffy; 110 Janos Marffy; 111 Gary Hincks; 112–113 Peter Sarson; 114–115 Pavel Kostal; 117 Janos Marffy; 120–121 Jonothan Potter; 124 Eugene Fleury; 125 Robin Jakeway; 126 Mainline Design; 127 Jonothan Potter; 127br Robin Jakeway; 128–129 Eugene Fleury; 130–131 Mainline Design; 131br Guy Smith; 132 Guy Smith; 134 Guy Smith; 138–139 Ann Winterbotham and Richard Draper.

Photographs

t = top; b = bottom; r = right; l = left; a = above; bw = below

8 Sinclair Stammers/Science Photo Library; 10 USDA/Science Photo Library; 12 Paddy Ryan/NHPA; 13–15 Kevin Schafer/NHPA; 20–21 Jim Amos/Science Photo Library; 28t Bryan & Cherry Alexander/NHPA; bl Jonathan Scott/Planet Earth Pictures; 28br John Kelly/The Image Bank; 29l Robert Harding Picture Library; 29r Penny Tweedie/Colorific!; 30 Peter Stephenson/Planet Earth Pictures; 32 Peter Scoones/Planet Earth Pictures; 51 G.I. Bernard/NHPA; 56 Jim Brandenburg/Planet Earth Pictures; 62 James Robinson/Oxford Scientific Films; 63 Stephen Downer/Oxford Scientific Films; 65 Deni Brown/Oxford Scientific Films; 68–69 Alastair Shay/Oxford Scientific Films; 72l Richard Coomber/Planet Earth Pictures; 72r E.A. Jones/NHPA; 73 Belinda Wright/Oxford Scientific Films; 74–75 Jean-Paul Ferrero/Ardea; 87 Patrick Fagot/NHPA; 95 ZEFA; 97 Space Frontiers/Planet Earth Pictures; 102 Frank Fournier/Contact/Colorific!; 103 Raghubir Singh/Colorific!; 108–109 Richard Hermann/Oxford Scientific Films; 110 Paul Berger/Tony Stone Images; 112 Peter Menzel/Science Photo Library; 116 James D. Watt/Planet Earth Pictures; 116–117 Roger Ressmeyer/Starlight/Science Photo Library; 118–119 Worldsat International/Science Photo Library; 121 Space Frontiers/Planet Earth Pictures; 122–123 Space Frontiers/Planet Earth Pictures; 123 B. Harris/ZEFA-Stockmarket; 124 Geoscience Features Pictures Library; 125 Warren Faidley/Oxford Scientific Films; 126 David A. Ponton/Planet Earth Pictures; 130 Warren Faidley/Oxford Scientific Films; 132–133 Warren Faidley/Oxford Scientific Films; 133 Pekka Parriainen/Science Photo Library; 134 Roger Ressmeyer/Starlight/Science Photo Library; 135l Rev. Ronald Rover/Science Photo Library; 135rt and 135ra John Sanford/Science Photo Library; 135rbw and 135rb John Bova/Science Photo Library; 137l Bryan and Cherry Alexander/Colorific!; 137r Steve Razzetti/Colorific!; 138 Mark Edwards/Still Pictures.